SLEIGHT OF MIND

SLEIGHT *OF* MIND

75 INGENIOUS PARADOXES IN MATHEMATICS, PHYSICS, AND PHILOSOPHY

MATT COOK

The MIT Press
Cambridge, Massachusetts
London, England

First MIT Press paperback edition, 2021
© 2020 Matt Cook

This book was set in Stone Serif and Stone Sans by Westchester Publishing Services. Printed and bound in the United States of America.

Library of Congress Cataloging-in-Publication Data

Names: Cook, Matt (Matthew), author.
Title: Sleight of mind : 75 ingenious paradoxes in mathematics, physics, and philosophy / Matt Cook.
Description: Cambridge, MA : MIT Press, 2020. | Includes bibliographical references and index.
Identifiers: LCCN 2019013954 | ISBN 9780262043465 (hardcover : alk. paper)— 9780262542296 (paperback)
Subjects: LCSH: Paradox.
Classification: LCC BC199.P2 C625 2019 | DDC 165--dc23
LC record available at https://lccn.loc.gov/2019013954

10 9 8 7 6 5 4 3

To Aristotle, the father of logic;
To Richard Feynman, who saved my father's life;
And to my father, whose curiosity burns brighter than the stars.

Contents

Acknowledgments

This book was a departure from my romance with thriller writing. Fiction is pure invention. The novelist creates the characters and joins in their adventures. Writing this book was an exercise in discovery. The process was a tale of many labyrinths, all part of a greater maze.

Inside a labyrinth, one sees but a few walls and corners. The complexity can be appreciated only from above. I am grateful to everyone who gave me lift when I needed the bird's-eye view.

Diana Zimmerman and Bob Dorian, my first steps through the maze began with you. As leaders of the Magic Castle Junior Society, you made it possible to understand the parallels between sleight of hand and sleight of mind.

Jim Spalding, you showed me the joy of mathematical paradox in high school. Your presentation of the Ross-Littlewood Paradox tempted me to buy a whole lot of tubing and tennis balls, but alas, no store in the country had the necessary inventory.

Nathaniel Sagman and Forte Shinko, you helped me comprehend several nuances of the Banach-Tarski Paradox. Thank you.

Aidan Chatwin-Davies, Michael Coughlin, and Nicholas Laurita, your passion for scientific education compelled you to take a risk on this book when it was hardly past the outline stage. Grant Sanderson, when asked to tackle a historically challenging question about the philosophy of mathematics, you asked: "When do I start?" Thanks to each of you for sharing your knowledge and contributing original material to this book. Our collaboration was a delight.

Alexandru Diaconescu, the maze would have been much longer without the benefit of your typesetting wizardry. I am grateful for your reliable help with problem-solving, and even more so for your friendship.

Victoria Skurnick, my literary agent, your wisdom has changed the way I view the pencil. I am now a better friend of the eraser. Thank you for helping me hone the craft of writing.

Jermey Matthews, many thanks to you and your colleagues at MIT Press for seeing something special in the connection between paradox and the magical arts. It is an honor to work with you.

Katja and Nicole Gisler, my sweet Swissters, you have given me the bird's-eye view not only of the Alps—from the cockpit of a ski plane at the "Top of Europe"—but also of the labyrinths of life.

Ross Askanazi, Vadim Kanofyev, Fidel Hernandez, Evan Miyazono, and Jon Zhang: You inspire me to look for labyrinths, treasure hunts, and mystery boxes. And when we're working through puzzles after midnight at the local diner, you inspire the dining staff to consider charging rent.

Members of my family, your support is the greatest constant in my life. You encourage me to ford every stream. From the bottom of my heart, thank you. My love for you is proof of Cantor's Paradox: There is no largest infinity.

Introduction

We've quips and quibbles heard in flocks,
But none to beat this paradox!
A paradox, a paradox,
A most ingenious paradox.
Ha, ha, ha, ha, ha, ha, ha, ha.
This paradox.
—Gilbert and Sullivan, *The Pirates of Penzance*

A paradox is really a sophisticated magic trick. A magician's purpose is to create the appearance of contradiction: to pull a rabbit from an empty hat, to violate the rules of logic and physics. Yet paradox doesn't require tangible things like rabbits or hats. Paradox works in the abstract, with words and symbols and concepts. It makes magic in the mind.

Most people can explain a magic trick using generalities. They may not know exactly how an illusion works, but they're familiar with the concepts of misdirection, sleight of hand, and gimmicks. Paradox achieves the same effect, but with nothing you can touch. There's no sleight of hand—only sleight of mind.

Paradox is the illusion of contradiction. A *paradox* is a set of two or more contradictory claims or arguments, each with the appearance of sound logic and reasoning. Reality allows no contradictions; paradox, as a concept, relies on the consciousness of a perceiver. A paradox creates the appearance of inconsistency in the universe, but it never identifies any genuine inconsistency. Any paradox can be resolved through the proper use of logic in dismantling the appearance of contradiction.

The question of whether there are any "real" paradoxes tends to conflate paradox with contradiction. The illusion of contradiction is a real cognitive phenomenon, but there are no real contradictions. Paradoxes do exist, just as magic tricks and other sources of mental discord exist. The existence of real paradox does not imply the existence of real contradiction, just as the existence of real magic tricks does not imply the existence of real magic.

The principle of noncontradiction (PNC) was first identified by Aristotle in *Metaphysics* Book IV (Gamma). Aristotle believed PNC to be the firmest axiom of the universe. According to Aristotle, noncontradiction is the axiom that makes reason, knowledge, and scientific inquiry possible. The principle, he said, applies to all fields and subjects, without exception. It is not a hypothesis that can be deduced logically or demonstrated experimentally, but rather an absolute and inescapable truth upon which deduction and demonstration rely, and a required presupposition for any rational philosophy. Aristotle believed that anyone claiming to reject PNC would necessarily surrender the capacity for intelligent thought, discourse, and action—and be condemned to a world of elusive truth and fallacious argument.

Aristotle was a thinker of extraordinary brilliance and prolific output, whose fields of study and interest encompassed philosophy, physics, geology, psychology, biology, medicine, psychology, art, language, and many other subjects. His teachings and writings have had a profound influence on civilization and rational thought. Among his achievements was the conception of formal logic. His ancient Greek predecessors passed down a rich intellectual heritage indeed, but logic was perhaps the greatest achievement in the championship of reason itself. Aristotle is often called the "father of logic" and a "father of Western philosophy."

The branch of philosophy concerned with developing a theory of knowledge is *epistemology*. Epistemology answers the question: What is knowledge, what makes it valid, and how do we gain it? To understand the relationship between our minds and our existence, it is important for us to define and understand the concepts of reason, rationality, and logic. The Russian-American philosopher Ayn Rand defines *reason* as "the faculty that identifies and integrates the material provided by man's senses" into concepts and abstractions. She considered reason to be one of man's three supreme values, along with purpose and self-esteem. Reason, she observed, is the method by which we learn the nature of the world and come to understand reality. She defines *rationality* as the virtue of recognizing and accepting reason as "one's only source of knowledge, one's only judge of values and one's only guide to action."

In his famous speech at the climax of Ayn Rand's magnum opus, *Atlas Shrugged*, John Galt defines logic in terms of its relationship to contradiction:

> Logic is the art of *non-contradictory identification*. A contradiction cannot exist. An atom is itself, and so is the universe; neither can contradict its own identity; nor can a part contradict the whole. No concept man forms is valid unless he integrates it without contradiction into the total sum of his knowledge. To arrive at a contradiction is to confess an error in one's thinking; to maintain a contradiction is to abdicate one's mind and to evict oneself from the realm of reality.

Among the corollary virtues of rationality are active-mindedness and intellectual ambition. *Active-mindedness* entails the critical evaluation of ideas and the willingness to update one's convictions—even core convictions—in accordance with logic and reason. *Intellectual ambition* is the desire to sharpen the mind: to expand one's rational faculty through a process of mental focus and exercise. The emotion that follows the embrace of these virtues is curiosity, the desire to learn and understand. Curiosity can also manifest itself as discomfort with apparent contradiction.

The purpose of this book is to provide a context for you to identify the rational process by which you resolve apparent contradiction. The forms of logic used in the resolutions may vary. With each example, try to classify the logic. Give it a name. Do you use mathematical principles? Formal logic? What kind of logic? Do you refine, expand, and generalize concepts that you have already formed? Do you create new concepts? Do you look for clarity and definition in ambiguous language? Do you challenge hidden assumptions? Asking these kinds of questions can make you a more effective learner and thinker. Resolving a paradox is one thing, but understanding the nature of a resolution is another. It's not just learning and thinking; it's learning about learning, and thinking about thinking.

Let's use an example to see what this kind of unpacking might look like. Consider two apparently contradictory claims: (1) Frederic has lived for twenty-one years, and (2) Frederic has had exactly five birthdays. This is precisely the "most ingenious paradox" referenced in Gilbert and Sullivan's comedic operetta *The Pirates of Penzance*, which tells the story of a boy named Frederic, who is contractually released from his indentures to a band of pirates upon his "twenty-first year." Frederic, it turns out, was born on February 29. The appearance of contradiction arises from the assumption that the same calendar date is reached every 365 days. Most people implicitly equate "number of years lived" with "number of calendar birthdays." This equivalence follows from the assumption, but the assumption is false. There are leap years. Case closed. Paradox resolved.

Or, is it? We can dig deeper. Why are there leap years? The time it takes the Earth to revolve once around the Sun is not an even multiple of the time it takes the Earth to rotate fully about its own axis. Why do we care to incorporate the Earth's rotational period into the calendar? So humans can time their activities around naturally evolved sleep cycles and navigate the world in the presence of light. It takes approximately 365.25 days for the Earth to complete one revolution. Without the leap year correction of February 29 every so often, calendar dates would no longer correspond to highly localized positions of the Earth's orbit around the Sun. In time, the seasons would be thrown off kilter. In 728 years, spring would become fall and summer would become winter. Why would that matter? There are benefits to synchronizing the calendar with expected temperatures and other seasonal changes. Among them has been the human ability to avoid starvation by optimally timing crop cycles.

The leap year was introduced over two millennia ago, in 46 B.C., by Julius Caesar, with the counsel of a trusted astronomer. The need for the leap year had been discovered even before that by the Egyptians. The "most ingenious paradox" is perhaps even more ingenious than Gilbert and Sullivan realized when writing their comedy. Frederic's plight is actually connected to agriculture and other aspects of civilization that have historically benefited from seasonal synchronization. The full resolution of the paradox calls for an understanding of calendar design and an investigation into the origin of the concepts "day," "season," and "year." Only by looking deeply into the relationships between these concepts and their value to human life could we see the full story behind why the initial assumption, that the same calendar date is necessarily reached every 365 days, is false.

This is the kind of resolution and analysis that you can expect in this book. Untangling the contradiction is half the story, and understanding the nature of the resolution—asking, "How did we just resolve that?"—is the other half.

The American logician and philosopher Willard Van Orman Quine categorized paradoxes into three classes: veridical, falsidical, and antinomical. A *veridical* paradox provides a sound argument demonstrating a result that seems absurd but is actually true. The "most ingenious paradox" is veridical because Frederic in fact has had only five birthdays despite his twenty-one years of life. A *falsidical* paradox provides an absurd result using logic that appears sound but is actually fallacious. A mathematical proof that sneaks in a forbidden division by zero can be used to prove that $1 = 0$; both the result and the proof are invalid, so such an example is falsidical. Finally, an *antinomical* paradox, more commonly called an *antinomy*, produces inconsistency using frameworks of reason that are considered standard and acceptable. Antinomies compel us to question and revise those frameworks. One famous antinomy is the Liar Paradox, a self-refutation that boils down to: "This statement is false."

The paradoxes will range broadly in difficulty and in the degree to which comfort with certain mathematical concepts may be helpful or necessary. The book can be enjoyed by intellectually ambitious readers of any background who calibrate the reading experience with their level of training and appetite for challenge. Readers are encouraged to focus on parts of the book that seem *just* out of reach. Typically, if a concept seems just out of reach, the effort of focus will make it comprehensible. That's how the learning happens. If you're tempted to skip part of a chapter because the concepts seem too foreign, you might try reviewing the introduction to that chapter. Each introduction is designed to serve as a crash course in the prerequisite concepts for the chapter. If the difficulty is still too great conceptually or notationally, there is no harm in moving to other parts of the book that you'll find more interesting.

The book is organized into chapters, each focusing on a certain class of paradox. The chapters are likewise divided into sections, each covering an individual paradox, although sometimes multiple related paradoxes are covered in a given section. Each section unravels with a presentation of the paradox at hand, a set of mutually inconsistent claims, and finally discussion and resolution designed to help you understand the source of mental discord and identify the cognitive process used to resolve it. Some paradoxes (sections) were difficult to slot into a single class (chapter), but a single decision was made for each based on the most pertinent concepts. At the end of the book, you will find a Notation Guide.

The chapters cover a wide variety of fields, including mathematics, logic, philosophy, the social sciences, and physics. The physics chapters were written by or with Aidan Chatwin-Davies, Michael Coughlin, and Nicholas Laurita. The final chapter includes a guest essay and poem by Grant Sanderson. I have welcomed the input of numerous peer reviewers and done my best to avoid mistakes, but some may have slipped through the cracks. Readers are encouraged to notify me of any errors discovered. An errata sheet will be maintained on my personal website, www.visitmatt.com.

Paradox is a feature of everyday life. It may not always be mathematical or otherwise academic in nature, but we encounter it with surprising, if not frustrating frequency, most often in the realm of human behavior. By reading this book, you've sought out even more paradox than what life presents on its own. You must have a special kind of curiosity. May these paradoxes delight that special curiosity of yours, dear reader, and show you something about the way your mind works.

1 Infinity

This view [of infinity], which I consider to be the sole correct one, is held by only a few. While possibly I am the very first in history to take this position so explicitly, with all of its logical consequences, I know for sure that I shall not be the last!

—Georg Cantor

Infinity is neither seen nor touched. It lies strictly outside the realm of perception and entirely within the realm of conception. There is no tangible infinity for man to observe with his senses. How, then, can we even speak of it? Recall that reason is the process by which we integrate concrete, sensory percepts into concepts, and the latter into yet higher abstractions. Reason is the faculty that permits us to conceive of intangibles like points, planes, liberty, and courage, none of which any human has ever seen, but which we can define and conceive.

Being so distant from the world of perceivable concretes, infinity is a challenging concept to grasp. We can much more easily imagine large but finite quantities, like the number of grains of sand on Earth, plankton in the oceans, or stars in the Milky Way. Infinity transcends our everyday understanding of numbers and arithmetic. The paradoxes in this chapter arise from counterintuitive properties of infinity. We begin with an introduction to some relevant concepts in set theory. Readers unfamiliar with set builder notation may reference the Notation Guide at the end of the book.

The German mathematician Georg Cantor, whose achievements include the invention set theory, proved something that is in itself paradoxical: Infinities come in different sizes. In fact, they come in an infinity of sizes.

Bizarre as it may seem, some infinities are larger than others, and for any infinity, there is always a larger one.

To prove or understand this discovery, we need a robust concept of size—a standard by which to measure the number of things in a collection. The concept must apply to both finite and infinite sets. Numerical quantity is good enough for finite sets; we simply count the elements. But numerical quantity, as a concept, is not powerful enough to handle infinities. We need a measuring tool that affords us greater power and versatility. It must extend the notion of size. That concept is *cardinality*.

Just as the ruler is a tangible tool used to evaluate length, the bijection is a conceptual tool that mathematicians use to evaluate cardinality. A *bijection* is a one-to-one correspondence between elements of two sets—that is, a way of pairing each element in each set to exactly one element in the other set, leaving no elements unpaired. If two sets can be bijected, they are said to have equivalent cardinality. Equal cardinality of two sets does not imply that every conceivable correspondence between them will be one-to-one. It just means that a bijection is possible. If no bijection is possible, then the sets have unequal cardinality.

The smallest infinite cardinal, designated *aleph-null* or *aleph-0* and denoted $\aleph_0$, is the number representing the size of the set of natural numbers. This set is said to be *countable* because if you were to start counting off the natural numbers one by one, any given number would be reached in finite time. That's not to say you would ever stop counting, but every natural number is a number you would eventually count off. The integers, counting numbers, odd numbers, even numbers, primes, and rationals are all countable for the same reason. They can be arranged in a list such that any given element will be reached or "counted" in finite time.

By definition, a bijection can be drawn between any two countably infinite sets, but it is not always obvious how to arrange such a bijection. Some sets may seem smaller than others but actually have the same cardinality. Take the counting numbers and the positive odd numbers, for instance. The counting numbers are $\{1, 2, 3, \ldots\}$ and the odd numbers are $\{1, 3, 5, \ldots\}$. The odd numbers are a subset of the counting numbers, so how could these two sets have the same cardinality? We can show that they do by lining up the two lists of numbers side by side and mapping the nth elements with each other:

$$1 \longleftrightarrow 1$$
$$2 \longleftrightarrow 3$$
$$3 \longleftrightarrow 5$$
$$4 \longleftrightarrow 7$$
$$5 \longleftrightarrow 9$$
$$\vdots$$

What if we wanted to biject the integers with the counting numbers? This seems harder to do because the counting numbers have a clear starting point and the integers go on endlessly in both the negative and positive directions. That's okay. We can either rearrange the counting numbers into a list without a starting point, or rearrange the integers into a list that does have a starting point and then line up the list of counting numbers with the list of integers. As a demonstration, let's use the latter method and arrange the integers into a list with a starting point, as follows (integers on the left, starting at zero):

$$0 \longleftrightarrow 1$$
$$-1 \longleftrightarrow 2$$
$$1 \longleftrightarrow 3$$
$$-2 \longleftrightarrow 4$$
$$2 \longleftrightarrow 5$$
$$\vdots$$

The fact that we can biject subsets and supersets is paradoxical in itself! The counting numbers, odd numbers, even numbers, primes, and integers can all be bijected with each other—or, by definition, with any other set that is also countable.

The union of any two countable infinities is still countable. To see this, imagine taking two countably infinite sets and enumerating their elements in two unending lists, A and B. Arrange the elements on each list so that each list has a starting point and continues forever. Create a new list, C, that enumerates the union of all listed elements in an alternating fashion. If we were to read the entries on list C one by one, we would still reach any given entry in finite time; and we could biject the elements in C with those in A or B as done previously. The union of A and B is still countable. It follows that the union of any finite number of countable sets is still countable.

Is the Cartesian product of n countable sets countable? The *Cartesian product* of sets is defined as the set of all n-tuples of elements that can be formed using elements from the original sets. In set-builder notation, that is:

$$X_1 \times X_2 \times \cdots \times X_n = \{ (x_1, x_2, \ldots, x_n) \mid x_i \in X_i \text{ for all } i = 1, 2, \ldots, n\}$$

A deck of 52 playing cards, for example, represents the *Cartesian product* of the set of suits and the set of ranks:

$$\text{Full deck} = \{\spadesuit, \heartsuit, \clubsuit, \diamondsuit\} \times \{A, 2, 3, 4, 5, 6, 7, 8, 9, 10, J, Q, K\}$$

What if we were to take the Cartesian product of a finite number of countably infinite sets? Our new set would still be countable. To see why, let us start with the Cartesian product of two countably infinite sets, A and B, and try to find a way to countably list the elements of their Cartesian product. Begin by arranging the elements of A and B into respective lists of enumerated elements, such that $a_i \in A$ and $b_i \in B$. Create a two-dimensional array with the A-list on the x-axis, starting with element a_1; and the B-list on the y-axis, starting with element b_1. Now generate a list of elements from $C = A \times B$ whose nth element is formed by pairing the coordinates of n as it appears here:

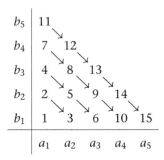

That is, the first element on your C-list is (a_1, b_1), the second element is (b_2, a_1), and so on, as you continue to list elements following the diagonals of your $A \times B$ array. Mission accomplished. Your new set C is countably infinite. By reading down the C-list, you will reach any given element in finite time because you will eventually reach every diagonal in finite time— and every diagonal contains a finite number of elements. It follows that the Cartesian product of any n number of countable infinities is still countable.

A larger infinite cardinal is the *cardinality of the continuum*. The contin- uum is the set of real numbers. The cardinal number of this set is denoted c or $2^{\aleph_0}$. Cantor showed through a proof by *diagonalization* that it is impossi- ble to make a complete, comprehensive list of all the real numbers. Hence, this infinity and all greater infinities are said to be "uncountable."

To understand Cantor's logic, imagine writing an unending list of all the irrational numbers in binary. Suppose that, after great, exhausting effort, you believe you have completed the list and included and enumerated every irrational number. Label each number by the order in which it appears on the list: Call the first number n_1, the second n_2, and so on. Now, define a function $f()$ that takes your ordered list and outputs a number whose nth digit is the opposite of the nth digit of the nth number on your list. Because the function outputs an irrational number that differs from every other irrational number on your list in at least some minor way, it must be that this new number is not on the list. The original list must not have been complete. The binary version of this argument extends to any other base.

Here's an example to illustrate the diagonalization argument:

$n_1 =$	$\boxed{0}$	0	0	0	0	0	0	0	0 ...
$n_2 =$	1	$\boxed{1}$	1	1	1	1	1	1	1 ...
$n_3 =$	0	1	$\boxed{0}$	1	0	1	0	1	0 ...
$n_4 =$	0	0	1	$\boxed{1}$	0	0	1	1	0 ...
$n_5 =$	1	1	0	1	$\boxed{1}$	0	1	1	0 ...
$n_6 =$	1	0	1	0	1	$\boxed{1}$	0	1	0 ...
$n_7 =$	1	1	1	1	1	0	$\boxed{0}$	0	0 ...
$n_8 =$	0	1	0	1	1	0	1	$\boxed{1}$	1 ...
$n_9 =$	0	1	0	0	1	0	0	0	$\boxed{1}$ ...
$\vdots$	$\vdots$	$\vdots$	$\vdots$	$\vdots$	$\vdots$	$\vdots$	$\vdots$	$\vdots$	$\vdots$

By applying the function $f()$ to this list, we obtain an output of 101000100... which by construction cannot appear on the original list, as it differs from every number n_i in at least the ith digit.

We learn from the diagonalization argument that it is impossible to cre- ate a one-to-one correspondence, or bijection, between the natural numbers and real numbers. There will always be real numbers left over, without partners. The set of real numbers is thus shown to have a greater cardinality. It is actually a larger infinity, and one that cannot be counted.

With these concepts under your belt, you are ready to take on the challenges of infinity. (An aside for more advanced readers: You might be wondering whether there is an *aleph-1* or $\aleph_1$, and if so, why it does not follow the discussion of $\aleph_0$. Indeed, there is such an aleph number. The meaning of $\aleph_1$ involves the concept of ordinal numbers, which is introduced in chapter 7. $\aleph_1$ is in fact the next largest infinite cardinal after $\aleph_0$. The question of whether $\aleph_1$ and c are the same is complicated. This is the subject of the *continuum hypothesis*, which states that there is no set whose cardinality is strictly between that of the integers and the real numbers. It has been proven that either this hypothesis or its negation can be added as an axiom to the most commonly used foundation of mathematics without contradiction. The truth of the continuum hypothesis is "undecidable." While $\aleph_1 = c$ under the continuum hypothesis, these numbers are still conceptually different.)

1.1   Hilbert's Grand Hotel

The following thought experiment was first posed by David Hilbert in a 1924 lecture. Imagine a hotel with a countably infinite number of rooms, numbered in ascending order (Room 1, Room 2, and so on). Every room in the hotel is occupied by a guest. A new prospective guest arrives at the concierge desk and asks if space is available. The clever concierge finds a way to accommodate the newcomer without causing any current guests to leave the hotel or share rooms: by asking each existing guest to move to the next room over. That is, the guest in Room 1 moves to Room 2; the guest in Room 2 moves to Room 3; and so on. Each current guest still has a room. Not only that, but Room 1 is now vacant, providing space for the newcomer. Did the concierge truly manage to create a vacancy in a fully booked hotel?

Claim 1

No. It is impossible to create a vacancy in a hotel if all the rooms are booked. The fact "All rooms are booked" implies that no newcomers can be accommodated, because otherwise, the number of guests would exceed the number of available spots. The concierge's trick would not work.

Claim 2

Yes. The trick devised by the concierge would indeed work. In a hotel with a countable infinity of rooms, "All rooms are booked" does not necessarily imply that "no newcomers can be accommodated."

Discussion and Resolution

Claim 2 is correct. There are three sets to consider: (1) the set of hotel rooms, (2) the set of current guests, and (3) the set that includes all the current guests and the newcomer. Each of these sets is countable and has the same aleph-0 cardinality. The infinities have the same size.

By definition, two sets have the same cardinality if and only if a bijection can be drawn between them. The bijection between the rooms and the current guests is given in the statement of the problem: Each guest has exactly one room, and each room has exactly one guest. When the newcomer arrives and each current guest gets bumped to the next room, a bijection is created between the hotel rooms and the set of all guests, including the newcomer. It is still the case that each guest has exactly one room and each room has exactly one guest.

The process can be repeated to accommodate a million, billion, or trillion new guests. Even more impressive, imagine a boatload of countably infinite prospective guests arrived. Could we accommodate them? In fact, we could. All the concierge has to do is free up a countably infinite sequence of rooms. There are infinite ways of doing this. One such way is to free up the odd-numbered rooms. He could do this by asking all the current guests to stand in a line and then, one at a time, in order, relocate to the next available even-numbered room. Every current guest would have a room, and the new boatload of guests could take the odd-numbered rooms.

What if there were a countably infinite number of such boatloads, comprising a fleet? What if there were a countably infinite number of such fleets, comprising a navy? The same logic would still apply because we'd still have a countably infinite set of people. This would not hold true if the nesting of infinities were itself infinite, for then the number of people would become uncountable. But, so long as the set of prospective guests is countably infinite, the concierge can accommodate all, despite the sign out front that says, "No Vacancy."

That's why the Grand Hotel consistently earns five stars from top rating sites. Be sure to book your visit soon. There'll always be room for you.

Closing Open Ends

The concept of "bumping" guests in an unending sequence can be used to create a bijection between a closed and an open set. Consider the task of bijecting $X = [0, 1]$ with $Y = [0, 1)$. The former is closed; the latter is open on the right. Begin by matching each pair of equivalent elements in X and Y. We almost have a one-to-one correspondence except for the fact that $1 \in X$ has no mate. No problem. Match $1 \in X$ with $\frac{1}{2} \in Y$, "bumping" the latter's previous mate. Now match $\frac{1}{2} \in X$ with $\frac{1}{4} \in Y$. Continue by matching every $\frac{1}{2^n} \in X$ with $\frac{1}{2^{n+1}} \in Y$. We are left with a one-to-one correspondence between the desired open and closed sets, having "bumped" elements along an infinite sequence of repositionings, much like the guests in Hilbert's Grand Hotel.

The same idea can be used to "close" a circle missing a point along its perimeter. Can you think of a simple way to fill a hole, leaving no new holes, by rearranging the existing points on the circle? You can do it like this: Pick any rational number d. Distance will be measured in a single rotational direction, clockwise or counterclockwise, along the curved path of the circle's circumference and starting from the missing point. Map the point that is a distance d along the circle's perimeter into your original hole. Map the point that is a distance $2d$ into the newly vacated spot, and so on, so that each point at a distance nd from the original hole maps to a point that is $(n - 1)d$ away. Because d is rational, it shares no common multiple with π. You will never run out of new points to fill the new holes you create because you will never land on the same point twice.

In each of these two examples, an open end is closed when all the points in an infinite series get bumped to a consecutive spot, much as the guests in the Grand Hotel get bumped to the next room. Each example describes only one of the uncountably many infinite series that can be used.

1.2 Hyperwebster

This thought experiment was constructed by Ian Stewart of the University of Warwick.

Suppose that an ambitious publishing company decides to print a dictionary so vast that it includes not only every English word, but every "word" that can be constructed using English letters, whether or not the word actually appears in the English vocabulary. Words appearing in this dictionary will therefore include ordinary vocabulary like PARADOX and SCIENCE, but will also include nonsensical strings of letters like XQZDL. The publishing company decides to include every possible word of every conceivable length, even words that have a countable infinity of letters. Because the dictionary is so vast, the company calls it the Hyperwebster.

The Hyperwebster would contain the entire works of Shakespeare and Chaucer. It would contain every book ever written and every conceivable alternative ending, including one in which Romeo and Juliet live happily ever after. It would contain your full biography and the biography of every one of your ancestors, a complete transcript of your life's thoughts, the solution to every Millennium Problem, a historically accurate account of the fall of the Roman Empire, and the correct explanation of the origin of the universe. It would be filled with a lot of nonsense, too. Every possible string of letters, even those of unending length, would appear in the Hyperwebster as a single word.

To publish these words in a dictionary, the company alphabetizes them in the standard way:

$$A, AA, AAA, ..., AB, ABA, ABAA, ..., AC, ..., AZ, AZA, ...$$

$$B, BA, BAA, ..., BB, BBA, BBAA, ..., BC, ..., BZ, BZA, ...$$

$$\vdots$$

$$Z, ZA, ZAA, ..., ZB, ZBA, ZBAA, ..., ZC, ..., ZZ, ZZA, ...$$

Realizing just how large this book would have to be, the publishing company decides to partition the Hyperwebster into twenty-six volumes—one for each starting letter. All the words starting with A belong in the A volume, all the words starting with B belong in the B volume, and so on. To avoid redundancy, the publisher removes the first letter of each word in each volume, because the first letter can be inferred from the volume title. The volumes now read:

Volume A: $A, AA, AAA, ..., B, BA, BAA, ..., C, ..., Z, ZA, ...$

Volume B: $A, AA, AAA, ..., B, BA, BAA, ..., C, ..., Z, ZA, ...$

$$\vdots$$

Volume C: $A, AA, AAA, ..., B, BA, BAA, ..., C, ..., Z, ZA, ...$

The publishing company realizes in shock that not only is every volume identical, but it contains the entire original Hyperwebster. In an attempt to decompose the Hyperwebster into twenty-six smaller books, the publishing company actually created twenty-six new copies of itself. An attempt at deconstruction yielded a miraculous feat of construction.

Claim 1

The publisher's ostensible achievement is logically impossible; it must be a fraud! Something cannot be created out of nothing. We can use a finite analogy to illustrate the point. Imagine deconstructing a building into its raw parts and materials. Organize those materials into different piles—the bolts, the wood planks, the concrete, and so on. Now try to replicate the original building repeatedly from each such pile. It can't be done. You do not have the proper building blocks. You cannot end up with more than the sum of the original parts. The same must be true of the infinite. The publisher's feat also violates the apparent hierarchy of sets. The Hyperwebster is the "parent" superset that contains the entirety of the "child" subsets (the volumes). A parent cannot be born from its child; a set cannot be hidden within its subset. Common sense tells us that the entirety of something cannot be hidden in one of its smaller parts. Somehow, a tricky manipulation must have been performed by the publisher.

Claim 2

Logic prevails over intuition. Believe it or not, the publishing company did in fact construct new copies of the Hyperwebster in an effort to decompose the original into distinct volumes.

Discussion and Resolution

Claim 2 is correct. The key insight, as in the case of Hilbert's Grand Hotel, is that *a subset of an infinite set can still have the same cardinality as the superset.* Intuition tells us that any proper subset must be smaller than its parent. This is true of finite sets, but not of infinite sets.

Although each volume is a subset of the original Hyperwebster, every set under examination is uncountably infinite. Sure, there are words in the Hyperwebster that do not appear in every volume; it may seem, then, that the volumes have fewer words. But the cardinality of each volume still matches that of the Hyperwebster.

Remember that we must add a degree of rigor to concepts like "bigger," "smaller," "more," "less," and "fewer" in the realm of infinity. We must employ a nuanced concept of size called *cardinality* and a special tool to evaluate cardinality called the *bijection*.

Decomposing the Hyperwebster into twenty-six volumes effectively created twenty-six unique subsets. Eliminating the first letter from each word in each volume created a bijection between each volume (subset) and the original Hyperwebster (superset). The Hyperwebster was not really hiding in each volume. Rather, by dropping the first letter in each volume, the publishing company discovered a simple one-to-one correspondence between the volume and the Hyperwebster.

Let us apply the same principles in a context that is easier to visualize. Imagine two sets, $A = [0, 2]$ and $B = [0, 1]$. That is, A is the closed set containing all real numbers from 0 to 2, and B is the closed subset of A that contains all the real numbers from 0 to 1. Our intuition tells us that set A must contain more points. After all, it is a superset of B. It contains every element in B and more—apparently twice as many. But can we create a bijection between A and B? The answer is yes, quite easily, by mapping every element n in B to the element $2n$ in A. You can visualize this by imagining shining rays of light as depicted between representative line segments (see figure 1.1).

The light rays pass through the line segment representing set B, mapping each point on B to a single point on the line segment representing set A (and vice versa). There is a one-to-one correspondence between the sets of points. Creating B as a subset of A is analogous to partitioning the Hyperwebster into volumes. Shining the flashlight as depicted is analogous to removing the first letter of each word in each volume. Both actions are tantamount to identifying a bijection between two uncountably infinite sets of the same cardinality.

1.3 Crossing Dimensions

We know that the set of real numbers has cardinality c, the *cardinality of the continuum*. The points along any line or line segment form a set of this cardinality.

Lines are one-dimensional. What if we studied objects of higher dimension? For example, how many points sit on the current page of this book?

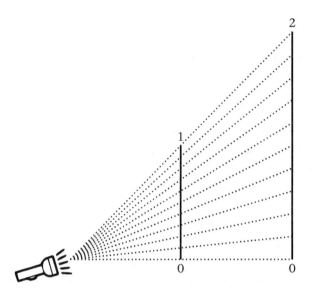

Figure 1.1
A bijection using rays of light.

How many unique coordinates can be assigned to the two-dimensional surface of this page?

Or, how about the number of points that lie within the Sun? How many unique coordinates can be assigned to the three-dimensional space occupied by the Sun?

Which of these numbers is the largest?

Claim 1

An object of higher dimension necessarily has more points, or unique coordinates contained within itself, than an object of lower dimension.

Claim 2

Every line, surface, three-dimensional area, or finite hyperspace in higher dimensions has the same number of points. The surface of this page contains just as many unique coordinates as the Sun.

Discussion and Resolution

Claim 2 is correct. A point is a theoretical construct with no dimension, whose primary characteristic is its location within a metric space. Points have no length, width, height, area, or volume. Still, it's difficult to see how a line could contain the same number of points as a surface, area, or higher-dimensional space.

Because we need to compare sizes of infinite sets, we need to invoke a familiar tool: the bijection. Can we create a bijection between sets of points contained within objects of different dimensions? Let's start by bijecting between the first and second dimensions, using the unit interval and the unit square as a case study. If this works, perhaps we can generalize to higher dimensions.

Let us start with the closed unit interval $A = [0, 1]$. Let us also consider the unit square $B = [0, 1] \times [0, 1]$. That is, B is a square of side-length 1 whose vertices exist at the origin and the points $(0, 1)$, $(1, 0)$, and $(1, 1)$. Our goal is to show that A and B have the same number of points: that we can create a bijection between all the coordinates on a line segment of length 1, and all the coordinates on a square of side-length 1.

Start by choosing a number on the unit interval that you wish to map to the unit square. Let's take the number 0.13009030067. Rather than viewing this decimal as a concatenation of digits, we're going to view this number as a concatenation of "molecules." Let's define a molecule as a nonzero digit and all of the zeros that precede it, going back to the last nonzero digit. Therefore, the molecules of 0.13009030067 are $m_1 = 1$, $m_2 = 3$, $m_3 = 009$, $m_4 = 03$, $m_5 = 006$, and $m_6 = 7$. The molecules are labeled by subscripts that specify their order.

Next, create two new decimals: the first by concatenating the odd-subscripted molecules, and the second by concatenating the even-subscripted molecules. Hence, from our example, we generate the decimals 0.13009006 and 0.3037. These two decimals are then paired to specify a coordinate on the unit square, which in this case is $(0.13009006, 0.3037)$. This method maps every point on the line A to a unique point on the square B, and vice versa. In a sense, we have managed to cross from the first dimension into the second!

Reaching even higher dimensions entails a straightforward generalization of the same process. Rather than constructing a pair of coordinates (x, y), use the same encoding method to generate n-tuples of coordinates for n-dimensional space.

Bijections between objects of different dimensions can also be made using *space-filling curves*. These are curves whose ranges can contain entire n-dimensional hypercubes, as constructed by taking limits of iterative bending and folding processes. In informal language, that means: By folding a line unendingly in special ways, you can eventually cover a surface or a higher-dimensional space.

1.4 Banach-Tarski Paradox

Can a pea be chopped into tiny pieces and reassembled into an object as large as the Sun? Practically speaking, no—but the idea may not be so outlandish in theory. This section presents an informal treatment of a paradox published in 1924 by Stefan Banach and Alfred Tarski, which violates geometric intuition in fascinating ways. Section 1.1, section 1.2, and the following introduction to groups contain important conceptual building blocks to this paradox.

An Introduction to Groups

An intimate familiarity with group theory and its vernacular will not be required to understand the Banach-Tarski Paradox, but a basic background may be helpful. Readers are encouraged to become comfortable with the following concepts before continuing to the explanation of the paradox.

A *group* is a set of elements, each associated with an action—like addition, rotation, reflection, and so on—such that any two elements can be operated together to form a third element. Hence, a group is a set G and an operation $\cdot$ that combines any two elements x and y in G to form another element in G, denoted $x \cdot y$ or xy. A group $(G, \cdot)$ has to satisfy the following basic axioms:

1. *Closure.* For any x and y in the group, $x \cdot y$ must also be in the group.

2. *Associativity.* For any elements x, y, and z in the group, $x \cdot (y \cdot z) = (x \cdot y) \cdot z$.

3. *Identity.* There is exactly one element in the group called the identity element, e, which can be thought of as doing nothing. That is, for all x in the group, $x = x \cdot e = e \cdot x$.

4. *Invertibility.* For every element x in the group, there is an element y in the group, which can be denoted as x^{-1}, such that $x \cdot y$ or $x \cdot x^{-1}$ equals e. That is, for any action x, the exact reverse can also be performed. In terms of consequences, performing an action and its reverse is the same as doing nothing.

A set of *generators* of a group is a set of elements whose actions can be performed repeatedly on themselves and each other to form all the elements in the group. The *free group* over some set *S* is the set of all strings that can be formed by concatenating the members of *S*, except trivially redundant ones that entail performing an action followed by its inverse.

The Rubik's Cube is an example of a tangible object that gives rise to a group. Manipulations of the faces of the mechanical puzzle are actions, the full set of which comprises the Rubik's Cube group.

The Paradox

Now, the fun part. The goal here is to start with a sphere and rearrange its points into two new spheres of the exact same size. Think that's impossible? Keep reading.

Let us start with a perfect sphere, sphere A (figures 1.2 and 1.3). Let's also give it two axes of rotation. These rotational axes don't necessarily have to be perpendicular, but to keep things simple, let's say they are. It's as if we've taken our sphere and pierced two skewers through it: one from the north pole to the south pole, and one through the equator, west to east. Hence,

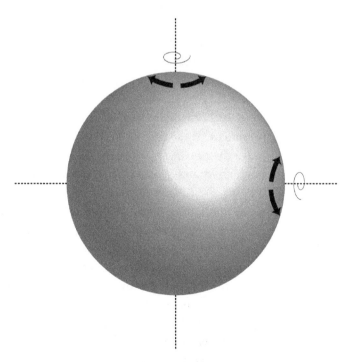

Figure 1.2
Sphere A, which has two axes of rotation.

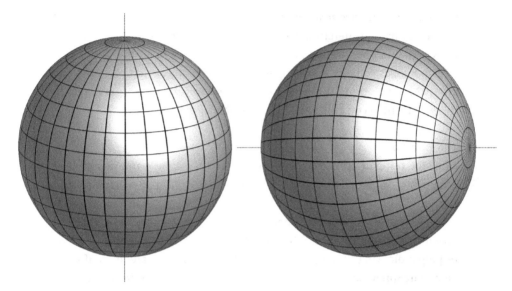

Figure 1.3
Sphere A shown twice, with lines of longitude and latitude relative to rotational axes.

the sphere has four poles. We can rotate the sphere around either axis by some fixed amount θ. For θ, we choose any value that is an irrational multiple of π radians, such as $\theta = arccos(\frac{1}{3})$, for a reason that will be explained shortly.

The rotations on sphere A comprise a set of actions. The sphere can be rotated up, down, left, or right. These four actions could be represented by two generators along with their inverses, collectively x, x^{-1}, y, and y^{-1}, but for ease of reference and understanding, we will denote these actions as U, D, R, and L. The pairs (U, D) and (L, R) are inverses; they cancel each other out if performed in succession, returning the sphere to the state it was in before either action was performed. In terms of consequences, performing an action and then its inverse is the same as doing nothing.

Our first objective is to name and partition all the rays that can be drawn from the center of the sphere, not including the center point, to a point on its surface. Every point in the sphere except the center point is contained on exactly one such ray. We will refer to a ray in terms of the point on the surface that it contains, and speak only of these points on the surface; the fact that we are referring to the corresponding rays is implied.

Let's begin by making a countably infinite list of every finite-length string that can be formed by concatenating the letters U, D, R, and L, excluding any strings that contain an action immediately followed by its inverse. For example, *DLUUR* will be on the list, but *DLRUU* will not because the latter contains a consecutive *LR* and so the string is equivalent to *DUU*, which is already on the list. Let's call this list our *superlist*. (In group theory terminology, this list is the *free group* formed by our rotational generators.)

Choose any arbitrary point on the surface of Sphere A. This point will be called an *origin* point. Using our superlist, we can now perform a countable infinity of finite actions on Sphere A, each designated by a string of rotations on the superlist. Doing so helps us obtain a countable infinity of points on the surface of the sphere by following the path of the origin point along the sphere's surface as each action is performed. Let's call this countably infinite set of points the *orbit* generated relative to its origin. Let's also name each point on the orbit. The name we'll give each point is the string of actions used to get to that point from its origin (figure 1.4).

Because θ is an irrational multiple of π radians, we will never reach the same point twice as we start naming points using our superlist of instructions—that is, unless the generated orbit contains any of the four poles. Why is the latter point true? Suppose we are on the north or south pole. Then, after a number of consecutive R rotations or L rotations, we are still on the same pole. The same goes for U and D rotations on the west-east poles. This will cause a small complication for us later, but for now, let's continue naming points on the surface of Sphere A.

Because the superlist is countably finite, the number of points named on the sphere relative to our origin is countably finite as well. There is an uncountable infinity of points on the sphere's surface, so we haven't even come close to naming all of the points yet. That's okay. We just choose a new origin from a point that hasn't been named, generate its orbit, and repeat this process an uncountably infinite number of times—until every single point on Sphere A belongs to some orbit and has a name in that orbit.

We can now partition Sphere A's points, across all orbits, into five subsets (*types*) and give each type its own color. Designate every origin point *black*. Designate every point whose name ends with L as *green*; call these *green* points or *left* points. Designate all *right* points *orange*, *down* points *blue*, and *up* points *red*.

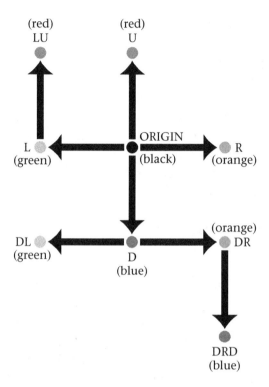

Figure 1.4
Naming and coloring points on an orbit.

The four orbits that contain any of the poles—let's call them the *polar orbits*—are going to cause a bit of trouble for us in the near term, so let's toss all points in those polar orbits into a hypothetical trash bin. Just delete them. We'll fill in the holes later. Let's put the point at the center of (inside) Sphere A into the trash bin and fill it later, too.

Imagine a hypothetical second sphere, Sphere B. It's just a concept right now. We haven't given it any substance. But it's ready for us, waiting to be filled in using only points from Sphere A.

Here's where the magic starts to happen. Let's take all the green *left* points and rotate them right, effectively adding an R to each name. This cancels out the terminating L in each green point's string; its new color is now determined by whatever letter, if any, came before the final L. Here's a small example of how points get renamed and recolored:

Former name	Former type	New name	New type
L	Left (green)	·	Origin (black)
DUL	Left (green)	*DU*	Up (red)
RDL	Left (green)	*RD*	Down (blue)
ULULL	Left (green)	*ULUL*	Left (green)

Of course, this renaming and recoloring applies to *all* of the uncountable green points; only four are listed here.

What have we accomplished? By merely rotating the green points, we have actually constructed the entire set of black, red, blue, and green ones! This is the essential innovation of the paradox. By moving an isolated piece of the sphere, we have somehow ended up with more pieces. Each rotated green point's new color is determined by the penultimate letter of its former name (or the terminal letter of its new name). Note that we haven't been able to construct any *right* (orange) points by rotating the green points right, because no string ending with an *L* was allowed to have *R* as a penultimate letter.

Let's drag our new black, red, and blue points into Sphere B, which we are building piece by piece. Here's the current status of our construction project:

Subset	Sphere A	Sphere B
Black	✓	✓
Red	✓	✓
Blue	✓	✓
Orange	✓	
Green	✓	
Polar orbits		
Center point		

Next, let's consider what would happen if we isolated the red *up* piece of Sphere A and rotated it down. We would end up with these pieces: red (*up*), green (*left*), right (*orange*), and black (*origins*). We would keep the red piece with Sphere A and move the new green and orange pieces to Sphere B, but then we would be left with a third black piece that we didn't need.

To prevent ourselves from constructing a third, unnecessary black piece of origins, let's rotate the red *up* piece down as we would have, but hold in place any point consisting entirely of consecutive *U*s. Rotating any such point down would have created a string with one fewer *U* and mapped point *U* to the origin. By holding in place any point labeled entirely by consecutive *U*s, we still end up with all the *n*-lengthed *U*-strings that we need without generating any new origins.

Having rotated almost the entire red piece down, we now have our original red piece, plus new green and orange pieces. Drag the new pieces to Sphere B, and let's reexamine the status of our construction project:

Subset	Sphere A	Sphere B
Black	✓	✓
Red	✓	✓
Blue	✓	✓
Orange	✓	✓
Green	✓	✓
Polar orbits		
Center point		

We are close! The only bits missing from each sphere are the polar orbits and a single center point. The good news is that the totality of these missing points is countably infinite, and we already have an uncountable infinity of points on each sphere that we can use to fill in the holes. We will do so using the principle employed in Hilbert's Grand Hotel (section 1.1). Construct a series of countably infinite points on each sphere. Let each point in the series be assigned the label of a counting number *n*. By moving each point *n* to the location of its predecessor, point $(n-1)$, every position in the series remains filled by a point, and we are left with an extra available point that was bumped out of first position in the series. By bumping out a countable infinity of such points, we can fill in the countable infinity of holes on each of Spheres A and B, to fill in the missing polar orbits and the center points.

Voilà. Our construction project is now complete:

Subset	Sphere A	Sphere B
Black	✓	✓
Red	✓	✓
Blue	✓	✓
Orange	✓	✓
Green	✓	✓
Polar orbits	✓	✓
Center point	✓	✓

Using only points from the original sphere, we have constructed a second sphere of the same size while maintaining the size and existence of our original. Can we repeat the process to create more and more spheres, both from Sphere A and its offshoots? Yes, we can. From a hypothetical star, we can construct a hypothetical galaxy. Or, from a hypothetical pea, we can construct a hypothetical Sun.

Claim 1

One should not buy into this Banach-Tarski tomfoolery. A sphere of fixed radius has a fixed and finite volume. Imagine filling a sphere with water to create a water balloon. By knowing the radius of the balloon, we can know exactly how much water it contains. A second water balloon can be created by neither practical nor theoretical means. Whether our sphere is tangible or hypothetical, filled with water or space or theoretical points, we cannot escape the logic that fixed volume is exactly that: fixed. The Banach-Tarski method must be flawed.

Claim 2

The steps that we've gone through here show us exactly how more spheres can be generated from an original, using only points from the original. It appears that more volume has been created, but we know that $1 + 1 \neq 1$, so we must revisit and refine our understanding of volume as a concept.

Discussion and Resolution

Claim 2 leads us down the correct path. We know that the set of points on one sphere, and the set of points on two or even n spheres, have the same cardinality. This may make the Banach-Tarski result less surprising at first. We can, in fact, biject the points contained in a pea with the points contained in the Sun, so why should it be interesting that we can create two spheres from one? The concepts underpinning this paradox are a bit more nuanced than bijections and cardinality.

The process of doubling the original sphere entails moving and rotating points and pieces without any warping, distorting, bending, or twisting of shape. How could mere movements and rotations serve to double the volume that the points comprise? The result violates geometric intuition. Our intuition applies to solid, measurable objects we see in everyday life. While the original Sphere A may be conceived as such a solid, the colored pieces are not solids as we usually think of them.

The concept that matters here is *measurability*. Every person has a basic intuition of what it means to measure something. We use measuring tapes and other devices to measure lengths in terms of inches, meters, and other units. We square these units to measure area and cube them to measure volume; or we use other units for these same concepts, such as hectares or acres for area, and gallons and liters for volume. Intuitively, these units of measurement apply to measurable concretes that we can see and touch with Euclidean geometry in one, two, or three dimensions.

But what, exactly, does it mean for something to be measurable, and how does the concept of measurement generalize to things we can't see and touch? Can the concept of measurement be extended to characterize mathematical constructs such as sets? These and other questions are explored by the branch of mathematics called *measure theory*. This theory says that the concept of measurement can be generalized, but sets must meet certain conditions to be measurable—that is, for the concept of measurement even to have meaning.

The formal proof of the Banach-Tarski Paradox relies critically on a postulate called the *axiom of choice*, which allows the construction of nonmeasurable sets. In the Banach-Tarski context, it is this axiom that allows us to make an uncountable number of decisions and take an uncountable number of actions to partition our sphere into pieces. By depending on this axiom, we are able to construct our colored pieces such that none of them actually has volume. That isn't to say that their volume is zero, but rather, the concept of volume does not even apply. (The definition of *measurability*, as well as a statement of the axiom of choice, are omitted.)

1.5 Cantor's Paradox

What is the largest infinity (cardinal number)? How big is it? We have seen from this chapter that there are infinities of different sizes. What do we get if we assemble a collection of all the infinities? How many different infinities are there?

Claim 1

There are two infinities: $\aleph_0$, which is countable, and $\mathfrak{c}$, which is uncountable. The greatest cardinal number is the larger of the two, $\mathfrak{c}$. This is the cardinality of the continuum, and it is the size of the set of all real numbers.

Claim 2

There are as many infinities as there are natural numbers. Just as there is no largest natural number, there is no largest infinity. The number of infinities is countably infinite, like the natural numbers.

Claim 3

There is no largest infinity or greatest cardinal number. Furthermore, it is impossible to form a set of all infinities or cardinal numbers. The vastness of all the infinities is so great that it transcends the very concept of size.

Discussion and Resolution

Claim 3 is correct. There is, in fact, no greatest infinity—but even more surprising, the "set of all infinities" is an invalid concept. There is no such set, because it would be so large as to transcend the concept of size.

To see why, we'll need the concept of a power set. The *power set* of some set X is the set of all subsets of X, including itself and the empty set. For example, if $X = \{1, 2, 3\}$, then the power set $\mathcal{P}(X)$ is

$$\mathcal{P}(X) = \{\emptyset, \{1\}, \{2\}, \{3\}, \{1, 2\}, \{1, 3\}, \{2, 3\}, \{1, 2, 3\}\}$$

For any finite set with n elements, its power set will contain 2^n elements. This is fairly easy to show. It is more difficult, however, to compare the size of an infinite set with its power set. Georg Cantor's famous theorem proves that for *all* sets X, even infinite ones, the power set of X is larger than X. Let vertical bars on either side of a set denote the cardinality of that set: $|X|$ is the cardinality of set X. These look just like absolute value bars but have a different meaning in this context. Cantor's theorem states that for all sets X, $|X| < |\mathcal{P}(X)|$.

Hence, the power set of any infinity will always be a larger infinity; for any cardinal number, there is always a greater one. So there is no largest infinity. This is not so counterintuitive, however. There is no largest natural number, but we can form the set of all natural numbers. Why can't we do the same for all cardinal numbers?

Suppose you could form the set of all infinite cardinals. List those cardinals in increasing order of magnitude. Next to each, define a set that has the corresponding cardinality, like this:

Cardinal	Set name
$\aleph_0$	S_1
$\aleph_1$	S_2
$\vdots$	$\vdots$

Now take the union of all the listed sets and call this T:

$$T = S_1 \cup S_2 \cup \ldots$$

It wouldn't really be possible to make such a list, but that's beside the point. The idea is that we're taking a union of sets such that every cardinality is represented by some set in that union.

Because T is the union of all the listed sets, the cardinality of any listed set can have at *most* the cardinality of T. Take the power set of T, $\mathcal{P}(T)$, which has a strictly greater cardinality than T. But because all infinite cardinals are listed (it's a set of all of them), the cardinal number $|\mathcal{P}(T)|$ should be on that list, which it is not. This is a contradiction, meaning we must reject the premise that the infinite cardinals form a set.

However big one might claim "all infinities" is, it's actually bigger and more infinite—so vast that it makes no sense to speak of "all infinities." It is unsettling to think that a collection could fail to qualify as a set, but the "set of all cardinals" is indeed a meaningless concept. The idea that some collections are invalid sets is explored in great depth in chapter 7. Advanced readers are encouraged to revisit this paradox after reading about the Burali-Forti Paradox (section 7.5). Can you find a connection?

Proof of Cantor's Theorem

Mathematically advanced readers may be interested in the proof of Cantor's theorem, which states that the cardinality of a power set is strictly greater than the cardinality of its corresponding set. Some definitions will be helpful. A mapping from set X to set Y is

- *injective* if every element of Y is mapped to by *at most* one element of X.
- *surjective* if every element of Y is mapped to by *at least* one element in X. It is possible for more than one element in X to map to the same element in Y.
- *bijective* if it is both injective and surjective.

$|Y| > |X|$ if and only if there exists an injection from X to Y, but not a surjection. X and Y have the same cardinality if and only if a bijection exists between them.

Now for the proof. This is a proof by contradiction: We will begin by assuming the opposite of what we want to prove, derive a contradiction, and thus show the initial assumption to be false. Suppose we have a set A whose power set is $\mathcal{P}(A)$. Assume that there exists a surjective mapping $f : A \to \mathcal{P}(A)$. This means every element of $\mathcal{P}(A)$, which (being the power set) is composed of subsets of A, is mapped to by at least one element in A.

Define B as a set of all elements in A that do not map to a subset of A containing that element. Formally,

$$B = \{x \in A \mid x \notin f(x)\}$$

Because $\mathcal{P}(A)$ is the set of all subsets of A and B is a subset of A, $B \in \mathcal{P}(A)$. Because f is surjective, there exists some y that maps to B:

$$\exists y \in A \text{ such that } f(y) = B$$

Is y a member of B? If y is not in B, then by the definition of B, y is in B. If y is in B, then by the definition of B, y is not in B. Here, y is a member of B if and only if it is not. Having reached a contradiction, we can now confidently reject the assumption that f is surjective. (The contradiction resembles Russell's Paradox, which is presented in chapter 7. The technique used to generate it is a form of Cantor diagonalization.)

There is no surjection from a set to its power set. An injection trivially exists from a set to its power set: map each element $x \in A$ to $\{x\}$ in $\mathcal{P}(A)$. These two conditions imply that the cardinality of a set is always strictly less than the cardinality of its power set.

2 Zeno's Paradoxes of Motion

Zeno's arguments, in some form, have afforded grounds for almost all theories of space and time and infinity which have been constructed from his time to our own.

—Bertrand Russell

Motion is a concept we form early in life. We observe it all around us all the time. Our lives require and imply motion. Motion even applies to things we cannot see, like particles, radiation, and spacetime itself.

Parmenides of Elea, often credited as the father of metaphysics and ontology (the taxonomy of concepts), was a pre-Socratic Greek philosopher and the founder of the Eleatic school of philosophy. His school disputed the validity of sensory experience as a means of acquiring knowledge and grasping truth, instead calling for purely logical standards. His poem *On Nature* made the case for two opposing views of reality: one according to the popular belief and the other according to logic. The senses, he stated, lead to popular opinions of the world that are false, whereas logic tells us of a world that is uniform, undifferentiated, timeless, and static.

Among Parmenides' students at the School of Elea was Zeno of Elea of circa 500 B.C., who defended the philosophy of his teacher by devising a series of paradoxes of motion. The paradoxes, he believed, proved motion to be impossible, despite our sensory evidence to the contrary. Zeno may have been the first to use the argument of *reductio ad absurdum*, or proof by contradiction. He was also credited by Aristotle as the inventor of the *dialectical method*, which uses reasoned discourse to resolve disagreement in pursuit of truth.

Figure 2.1
Homer dashing along a partitioned track in either direction.

Parmenides was right to point out the difference between truth and appearance. His mistake, however, was in dismissing sensory perception altogether, while relying on it to form the concepts his argument employed. This mistake resembles a fallacy called the *stolen concept*, a logical error identified and named by Ayn Rand. The stolen concept fallacy is committed by using a concept as a pillar, ingredient, or presupposition in an argument that seeks to invalidate the same concept. An argument cannot depend logically on that which it seeks to negate. Parmenides and Zeno depended on their senses to form concepts with which they would argue that senses cannot be trusted. At a practical level, they relied on the possibility of change and motion to communicate the ideas that change and motion are impossible.

The paradoxes themselves, however, commit fallacies that are more subtle and difficult to untangle. They remain profound and at times disconcerting, even thousands of years after their conception.

2.1   Dichotomy

Suppose Homer is on a linear racecourse of length 1, which he begins traversing when a whip cracks (figure 2.1). To reach the finish line, he must first traverse half the course, or a length of $\frac{1}{2}$. Then, he must traverse half of the remaining course, or a length of $\frac{1}{4}$, after which he must do it over and over, ad infinitum. Each time Homer reaches a former midpoint, he creates a new midpoint between himself and the finish line, which he must cross before reaching the finish line. There can be no final midpoint in the countably infinite series of midpoints. To cross the finish line, Homer must accomplish an infinite number of tasks in finite time, which Zeno argues is impossible. Furthermore, there is no final step.

The argument also works in reverse, with no first step. For Homer to reach the end of the course, he must get halfway there. To get halfway there, he must get a quarter of the way there, and an eighth of the way there, and a sixteenth of the way there, and hence follows the infinite regress. Either argument, Zeno holds, proves motion to be impossible because traversing any distance of positive length requires the impossibility of performing infinitely many tasks in finite time.

Claim 1

Zeno's logic prevails over the senses. Parmenides and Zeno were right: Physical motion must be impossible. We live in a static, motionless world. Not even time itself can progress.

Claim 2

We know motion to be possible—we observe it all the time. The simple act of reading this line required movement of the eyeballs. Zeno's logic must be flawed.

Discussion and Resolution

Claim 2 is correct, but there are a variety of treatments of the resolution, each of which addresses a different aspect of the paradox.

One treatment addresses the following not-so-subtle question: Do infinitely many line segments of finite length necessarily span an infinite length? We can prove the answer to be *no* as follows. For some $x \in (0, 1)$, let z be the infinite sum

$$z = \sum_{n=0}^{\infty} x^n = 1 + x + x^2 + x^3 \ldots$$

Multiply both sides by x to obtain

$$zx = 1(x) + x(x) + x^2(x) + x^3(x) \ldots = x + x^2 + x^3 + x^4 \ldots = z - 1$$

Solving for z in $zx = z - 1$, we obtain

$$z = \frac{1}{1 - x} \tag{2.1}$$

The infinite sum z is finite for any positive x that is less than 1. We can apply this result to the context of Zeno's Dichotomy by plugging in $x = \frac{1}{2}$, which yields a series of operations such that each successive operation brings us half the distance to the finish line.

We have to be careful, though, when we speak of performing an infinite number of operations. What does it mean to add quantities an infinite number of times? The arithmetic shown here may have yielded a finite result, but the construction of the problem assumed away the true essence of Zeno's Dichotomy: the issue of performing an infinite number of operations in finite time. We defined z as an infinite sum, but are we even allowed to take an infinite sum? Does time allow it? Does an infinite sum even pass the test of definition?

One way of approaching this issue is to consider the quantity 2^{-n} as the duration of time it takes to traverse each corresponding distance while moving at a constant speed. The operation of addition takes no time; here, addition is an abstract function that we can imagine the universe performing instantaneously as time flows. We need not concern ourselves, then, with the fact of the infinite quantity of operations that the universe inevitably allows, but rather with the sum of durations—and we have shown that sum to be finite.

A finite-span time or distance can be dissected into an infinite quantity of finite spans. Putting the individual parts together requires an infinite number of operations, but we can think of nature as performing them—inevitably, inescapably, and unrelentingly—as Homer runs his course. The operations themselves take no time; they are mere abstractions, ways of relating the racecourse to a summation.

Mathematically, we resolve the issue of executing infinite operations by employing two useful concepts: *convergence* and *limits*. These concepts were developed by the French mathematician and physicist Baron Augustin-Louis Cauchy, whose pioneering work spanned real and complex analysis, abstract algebra, calculus, and much more.

In plain terms, a sequence of numbers is said to *converge* to a *limit* if the sequence's progression becomes arbitrarily close to that limit. In more formal language, a sequence $X = \{x_1, x_2, \ldots, x_\infty\}$ is said to converge to some limit L if for every tiny number ε, there exists an i such that for all $n \geq i$, x_n is within a distance ε of L. In plain English, however tiny a distance you could name, the series will eventually stay within that distance of its limit.

Limits and convergence, as concepts, help us assign numerical value to a series (when possible), called the *sum of the series*. The sequence of locational milestones describing Homer's journey can be represented by the sequence of partial sums of the individual distances traversed:

$$\left\{ \frac{1}{2}, \frac{3}{4}, \frac{7}{8}, \frac{15}{16}, \dots, \frac{2^n - 1}{2^n}, \dots \right\},$$

which can be shown to converge to a limiting value of 1. What that means is that however small a number we could pick—even the reciprocal of a googolplex—there is some element in our sequence that is within that distance of 1, as are all of that element's successors. The limit of Homer's course is not some infinitely far-off point, but rather the end of the racetrack.

There remains one more issue to address with Zeno's Dichotomy, and that is the problem of having no initial or terminal step. Let us be specific about what is meant by a step, as it could mean one of two things: moving a positive distance or attaining a point along the racetrack. Because the distances covered by a step converge to zero, we will use the latter definition. Homer's supposedly impossible task is to attain a countably infinite number of points between the start and the finish line, which can be constructed in such a way as to have no beginning and/or no ending.

Let's take things a step further to make it harder for Homer. The two sequences Zeno identified are only two of uncountably many countable sequences that can be formed using the continuum of points between the starting point and the finish line. Any such sequence might have no beginning and/or end. Or we could take the most pessimistic view: Homer has to traverse every point along the way, a continuum of points, uncountably many, with no beginning and no end if the start and finish lines are excluded.

This problem is purely psychological. Any discomfort is based on the premise that to move at all, Homer must take a first step (attain a first point), and that to finish, he must take a last step (attain a last point). That premise is false—and to see why, we must revisit the concept of motion.

Motion is intuitively conceived through the lens of velocity, as the change in an object's location over time. An object can move from one location to another only by passing through intervening points at some nonzero velocity, and therefore it is the property of having nonzero velocity that implies and defines motion to most people. That concept of motion may be good enough for everyday life, but to fully resolve Zeno's Paradoxes, we must scratch this notion of motion and replace it with a more nuanced and helpful formulation.

The more useful concept, and the one used in modern physics, is that *motion is a functional relationship between instants of time and points in space.* Motion is conceived as a continuous correspondence between a set of instantaneous moments and a set of coordinates indicating position. Every point in time must correspond to exactly one point in space. It is this concept of motion that permits the proper concept of velocity as the first derivative of the motion function. Velocity emerges from the motion function, but it is not the essence of motion.

Properly viewed, Homer's racecourse is characterized by a measurable continuum of points in space. The race's duration is characterized by a measurable continuum of points in time. Homer's motion is the continuous correspondence between the time set and the position set.

This concept of motion helps us understand that motion really doesn't require a first step (point) or a last step (point). The fact that the racetrack between the starting and finish lines is an open set poses no problem. All that matters is the existence of a functional mapping between the time set and the location set.

It is interesting to note that at the point in time when the starting whip cracks (at which time Homer is on the starting line), Homer's motion function must be infinitely differentiable from the right. Starting from zero, a change in location implies a change in velocity, which implies a change in acceleration, which applies a change in jerk, and so on.

The concepts of convergence, limits, motion, and velocity as formulated in this section will continue to serve us as we visit Zeno's remaining paradoxes. In fact, if you've made it through the Dichotomy with clear understanding, the others should be straightforward.

2.2 Achilles and the Tortoise

Suppose Achilles, the swiftest of Greek heroes, and a tortoise have entered into a race (figure 2.2). Because Achilles runs much faster than the tortoise, the tortoise is given a head start. The race begins. By the time Achilles reaches the tortoise's starting point, the tortoise has advanced some distance, however small, to a new milestone. Achilles keeps running. Once again, by the time Achilles reaches the tortoise's previous milestone, the tortoise has advanced to a new one. The logic continues into infinite regress: Although Achilles runs faster and continues to close in on the tortoise, Zeno argues that Achilles can never overtake the tortoise because doing so would require the completion of an infinite number of tasks and the traversal of an infinite number of positive distances to an infinite number of milestones.

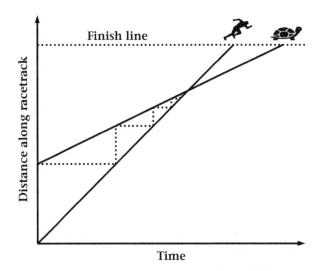

Figure 2.2
Achilles chasing the tortoise, reaching various milestones.

Claim 1

It appears that Achilles' heel is but one of his two vulnerabilities—the tortoise is his second. Zeno's logic was sound: Achilles will never overtake the tortoise.

Claim 2

Zeno's logic is flawed. The equation *distance* = *rate* × *time* can help us calculate exactly when Achilles will overtake the tortoise, and we can debunk Zeno's argument using concepts we have already identified.

Discussion and Resolution

Claim 2 is correct. This paradox is analagous to Zeno's Dichotomy (section 2.1), so its treatment is brief.

The concepts of convergence and limits, as well as our refined concept of motion, apply here. The paradox is resolved by showing that the sequence of partial sums used to compute the locations of the milestones converges to a point at which the two rivals meet. Achilles then proceeds to surpass the tortoise. The problem—that Achilles faces no terminal milestone—is handled by our new concepts of motion, as presented in section 2.1.

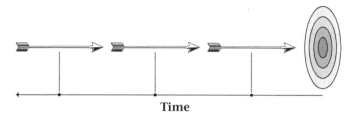

Figure 2.3
Arrows at rest at every point in time.

2.3   The Arrow

Zeno argued that for an object to be in motion, it must change its location over the course of time. He used an example of an arrow soaring toward a target. At any instantaneous moment in time, the arrow is moving neither into nor out of its current location in space toward its destination. If it were to do so, it would occupy a space greater than itself in order to have somewhere to go. Furthermore, if an object were to move during an instantaneous moment, we would be able to divide that moment into parts, and yet we have required our moments to be indivisible by definition. Therefore, at any point in time, the arrow is immobile. Time is composed entirely of instantaneous moments of zero duration, and at every one of them, the arrow is at rest (figure 2.3). Therefore, the arrow is in a permanent state of immobility. Motion is impossible to the arrow, and to everything else.

Claim 1

Sign me up to the Eleatic School. Zeno has proved that my senses are deceiving me. The arrow, and the universe as a whole, must indeed exist in a permanent state of immobility. Every object is uniformly at rest. Motion is impossible.

Claim 2

Zeno's argument is flawed and can be debunked using our new, sophisticated concepts of motion and velocity.

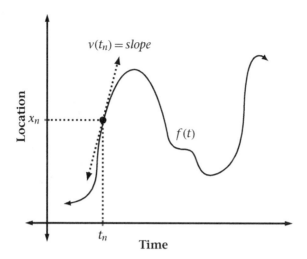

Figure 2.4
A motion function in one dimension.

Discussion and Resolution

Claim 2 is correct. Calculus gives us the tools to penetrate the Arrow Paradox. Recall our new and improved notion of motion: *Motion is a functional relationship between instants of time and points in space.* Motion must be treated as a continuous mapping between a set of time points and space points. Motion is not the property of having nonzero velocity. Velocity is defined as the rate of change of the motion function, calculated by taking the first derivative. It *is* valid to speak of an object's velocity at a given point in time, but to speak of motion at a point in time is to confuse the concepts. At any given instant in time, Zeno's arrow does have velocity equal to the slope of the motion function at that instant.

Figure 2.4 illustrates the functional relationship between space and time as a concept of motion. The x-axis represents the time continuum, and the y-axis represents a one-dimensional space continuum. The function $f(t)$ continuously maps each point in time to a point in space. The slope of the line tangent to $f(t)$ at t_n gives us the velocity of the moving object at time t_n.

Note that we have not invoked any theorems or formal proofs in calculus to resolve the Arrow Paradox. We have revisited and refined basic abstract concepts that people form as children. Concepts are mental classifications of things that have similar characteristics. We form concepts because they

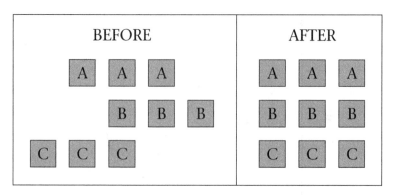

Figure 2.5
Zeno's Stadium.

are useful to us; they help us process the world around us. As we discover, by means of our senses, new characteristics of concrete things—or as we discover, by means of logic, new characteristics of abstractions—reason compels us to revise our formulations of these concepts or create new concepts altogether.

It has been through this process of conceptual refinement, reformulation, and creation that we have resolved Zeno's paradoxes. We have determined that summation cannot sufficiently handle an infinity of operations, so we have created (invoking Cauchy's seminal work) concepts of convergence and limits. We have determined that the intuitive concept of motion cannot sufficiently describe the real world, so we have reformulated it into one that does.

2.4 The Stadium

Suppose we have three rows of an equal number of similar objects. The objects are slotted into vertical alighnment as if they also belong to columns. The rows are labeled *A*, *B*, and *C*. Row *A* will remain motionless. Now suppose we simultaneously shift row *B* one slot to the left and row *C* one slot to the right (figure 2.5). An object in row *B* passes one unit in row *A*, but two units in row *C*. Rows *B* and *C* were moving just as fast as each other, and yet different lengths were traversed. Due to the relationship between distance, velocity, and time, we can see that one positive unit of time, and half the same unit of time, must be equal.

Claim 1

Zeno was right. For some positive duration of time T, $T = \frac{T}{2}$.

Claim 2

Zeno did not understand that speed is relative.

Discussion and Resolution

This paradox is rather easy to untangle. The story of the moving rows suggests that Zeno was thoroughly confused about the relative nature of velocity. He seemed to treat velocity as an absolute—a property independent of any frame of reference. The fact that row B moves twice as fast relative to row A as it does relative to row C is not a difficult reality to accept. We can compare the relative velocities of different pairings of rows without violating any principles of nature—and frankly, without challenging our modern intuition of how things move.

Nonetheless, the Stadium does pose an interesting thought experiment if viewed through a different lens. Imagine a discrete, quantized universe in which space and time were comprised of positive-sized pockets called *atoms*—tiny, indivisible units just like the pixels on a digital image. In such

Figure 2.6
A pixelated game avatar.

a world, it would be possible for entities occupying these pockets to pass each other without ever aligning. An object in row B could pass an object in row C and never exist side by side. This is probably not a hard conceptual pill to swallow, either (particularly for fans of old video games with highly pixelated graphics, like the avatar in figure 2.6).

To physicists, the potential for a quantized universe is more than a thought experiment. It is a topic of intense study. Zeno's Paradoxes may be resolved, but investigation persists into the nature of spacetime as discrete or continuous. If the universe is in fact made of positive-sized, indivisible units, then under some very powerful microscope, our bodies and environments might look and behave just like characters and worlds in a video game.

3 Supertasks

My greatest fear: repetition.
—Max Frisch

A *supertask* is a philosophical concept referring to a countably infinite sequence of operations performed within finite time. Philosopher James F. Thomson coined the term *supertask* in 1954 believing, like Zeno, that an infinity of operations could not be performed in finite time. As we have seen, Zeno's Dichotomy and his paradox of Achilles and the Tortoise both entail supertasks. What follows is a further sample of paradoxical supertasks selected from the many conceived by mathematicians, physicists, and philosophers.

It is important in this chapter to distinguish practical, physical impossibility from theoretical, logical impossibility. The supertasks presented are all practically impossible to perform. Performing these tasks would require moving faster than light or breaking an object's continuity in its path through spacetime, or completing operations within the duration of a Planck time unit—the time it takes light to travel the incredibly small Planck length (named after Max Planck), at which scale it is theorized that classical notions of gravity and spacetime fail and quantum effects take over. Physicist Gustavo Romero (2014) has argued that any attempt to carry out certain kinds of supertasks "will produce a divergence of the curvature of spacetime, resulting in the formation of a black hole." Romero maintains that "supertasks, contrarily to popular view among philosophers, are physically impossible. Supertasks, literally, collapse under their own weight."

Despite their practical difficulties, supertasks can still be imagined in ways that appear paradoxical in pure theory. Unless otherwise stated, our purpose here is to untangle apparent contradictions in the theoretical realm.

3.1 Thomson's Lamp

Suppose we have a lamp starting in the *on* position at time $t = 0$ minutes. After a half-minute, we switch the lamp off. After another quarter-minute, we switch it on, and after another eighth of a minute, we switch it off again. Each time we get halfway closer to the 1-minute mark, we change the state of the switch. The process continues ad infinitum. Time moves inevitably onward until we reach $t = 1$. At this time, is the light on or off? Thomson argues that it cannot be on because every *on* action was followed by an *off* action. Conversely, the lamp cannot be off because every *off* action was followed by an *on* action. The lamp can be neither on nor off, but there is no other state for it to have.

Claim 1
Thomson must have been right: It is impossible to perform an infinite number of options in finite time. By accepting impossibility as a premise, we have allowed the construction of an absurd thought experiment—one that can never be performed in reality, and therein lies the resolution.

Claim 2
The impossibility of actually performing this supertask in real life does not bar us from addressing it logically in the abstract. The resolution of this paradox lies in separating what we can and can't deduce from the lamp's history. We can't be expected to describe something that is not deducible, and therefore the indeterminate status of the lamp should not bother us.

Discussion and Resolution
Claim 2 is correct. To see why, let's begin by using all the information we are given to construct a functional relationship between points in time and the state of the lamp. By construction of the problem, the domain of our function $f()$ is the open set of times $T = [0, 1)$, and the codomain is the discrete set of states $S = \{ON, OFF\}$. Our function maps the domain into the codomain:

$$f:T\to S$$

We construct the function as follows. Partition the set T into countably infinite subsets of time intervals. Each time interval starts with a state change and is open on the right, including all points of time before the next state change. Let the series of subsets be $\{T_0, T_1, T_2, ..., T_i, ..., T_\infty\}$, where the subscript denotes the ith state change, assuming ON at time zero. For example, $T_0 = [0, \frac{1}{2})$, $T_1 = [\frac{1}{2}, \frac{3}{4})$, and so on. For any integer $i > 0$, we can calculate T_i as

$$T_i = \left[\frac{2^i - 1}{2^i}, \frac{2^{i+1} - 1}{2^{i+1}}\right) \text{ for all } i = 1, 2, ..., \infty$$

We can now specify the function f for some time $t \in T_i$:

$$f(t) = \begin{cases} ON, & i \text{ is even} \\ OFF, & i \text{ is odd} \end{cases}$$

Because the union of subsets T_i is T, the function maps every desired point in time into a state of the lamp.

Let's return now to the original question: What is the state of the lamp at $t = 1$? What is the value of $f(1)$? We cannot answer that question, nor should we be troubled by our inability to answer it. Here, 1 is not an element in the domain T, and so our function does not apply. We constructed the function with all of the information we were given, deducing the state of the lamp at a continuum of points, but that time continuum does not include $t = 1$ or beyond. For all we know, the lamp no longer exists at $t = 1$. Or perhaps it has been replaced by a mythical creature from Pluto. We simply do not know anything about the lamp at $t = 1$.

Paul Benacerraf pointed out in 1962 the importance of asking whether the history of states enables us to logically determine the state at $t = 1$. He concluded that it does not. Because time moves inevitably forward, it may *feel* reasonable to ask what the function outputs at $t = 1$. But our function, as determined by the statement of the problem, does not include $t = 1$ in its domain.

Suppose you had a function that mapped *letters* to *numbers*. If you input a *color* into such a function, would you really expect an answer? Would you even find the impossibility of an answer disconcerting? Alternatively, if you were told about a hypothetical curtain and nothing else, would you be troubled by your inability to deduce what was behind that curtain? Would you feel like you somehow should know what was behind it?

The psychological discord arises from the intuitive but false assumption that the state of the lamp should be deducible, even though we are told nothing about its state at $t = 1$. We should feel fully satisfied by our ability to create a functional relationship between T and S, even if $1 \notin T$.

3.2   Ross-Littlewood Paradox

The following thought experiment was published in J. E. Littlewood's book *A Mathematician's Miscellany* in 1953.

Suppose you have a very long tube and a countable infinity of balls labeled 1, 2, 3, ad infinitum (figure 3.1). At time $t = 0$ minutes, the tube is empty, and you start the clock. After a half-minute, you insert balls labeled $1 - 10$ into the tube from the right, in numerically increasing order, and immediately remove the leftmost ball (ball 1). After another quarter-minute, you insert balls labeled $11 - 20$ into the tube, in numerically increasing order, and immediately remove the leftmost ball (ball 2). Each time you get half the distance to the minute mark, you add the next ten balls and remove the leftmost ball (see figures 3.1 and 3.2). Soon, a flurry of motion ensues, until at last the timer reads $t = 1$. How many balls are in the tube?

Claim 1

There must be an infinite number of balls in the tube. Each action adds nine total balls. After n actions, there must be $9n$ balls in the tube. As n approaches infinity, so does $9n$. The number of balls in the tube keeps getting larger, without bound.

Figure 3.1
Numbered balls may pass in and out of a tube.

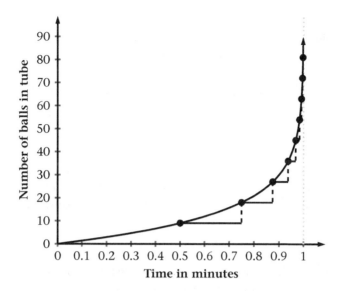

Figure 3.2
Number of balls in the tube as time proceeds, with a fitted curve.

Claim 2

There must be zero balls in the tube. For any given ball that can be named by number, the exact time of its removal can be calculated.

Discussion and Resolution

As described, neither claim is technically true *yet*, but by making certain assumptions and studying certain variations of the problem, we can construct scenarios that will lead us toward one claim or the other, or various other truthful claims. We will begin by studying the original variation as presented.

Recall from Thomson's Lamp that the series of steps preceding $t = 1$ does not logically imply a state of the lamp at $t = 1$. We are told only of the state of the lamp over the interval $T = [0, 1)$, which is open on the right.

With respect to Ross-Littlewood, we have been given information about a process by which balls are inserted and removed from a tube over the same time interval $T = [0, 1)$. This information can be used to deduce the state of any given ball as *IN* or *OUT* at any time $t \in T$, but we have been given no information that logically implies the state of any given ball at $t = 1$. Psychologically, it *feels* like the problem's setup has logical implications with respect to the state of each individual ball, and the tube as a whole, after the minute is up. This is not the case. As far as we know, the tube may have beamed to another planet at $t = 1$ and been replaced by fairies.

If this explanation is psychologically difficult to accept, it may be helpful to identify the functions, formally and rigorously, governing the state of each ball, which collectively govern the contents of the tube. We will use all the information we have been given.

Let us begin with some notation. Let $B = \{b_1, b_2, \ldots, b_i, \ldots, b_\infty\}$ denote the set of numbered balls. Let $S = \{IN, OUT\}$ be the set of possible states that a ball can have. Let f be a function that takes as inputs a particular ball and a particular point in time, and outputs the following state:

$$f : B \times T \to S$$

The task is now to construct f. We know that every ball b_i starts OUT of the tube at $t = 0$, remains OUT until it is put IN, and remains IN until it is removed. Here, f must incorporate the timing of these events. Let's break these steps into component functions.

Let $\lceil x \rceil$ denote the *ceiling* of some number x; that is, x rounded up to the nearest integer. Let t_{IN} be a function that outputs the time at which a given ball is first inserted into the tube. This function, therefore, is a mapping from B to T:

$$t_{IN} : B \to T$$

We can calculate the t_{IN} function as

$$t_{IN}(b_i) = \frac{2^{\left\lceil \frac{i}{10} \right\rceil} - 1}{2^{\left\lceil \frac{i}{10} \right\rceil}}$$

Now we need a t_{OUT} function that tells us the point in time at which a given ball b_i is removed from the tube. Such a function has a similar structure:

$$t_{OUT} : B \to T$$

We can calculate the t_{OUT} function as

$$t_{OUT}(b_i) = \frac{2^i - 1}{2^i}$$

With these component functions, we are ready to compile our piecewise f function:

$$f(b_i, t) = \begin{cases} OUT, & t \in [1, t_{IN}(b_i)) \\ IN, & t \in [t_{IN}(b_i), t_{OUT}(b_i)) \\ OUT, & t \in [t_{OUT}(b_i), 1) \end{cases}$$

We have incorporated into our function the fact that after a ball is removed, it stays out of the tube for the remainder of the time domain. Mathematically, this fact is stated this way: For all $b_i \in B$ and $t \in [t_{OUT}(b_i), 1)$, $f(b_i, t) = OUT$. This implies that the function is continuous on the rightmost part of the domain and converges to the limiting value of the OUT state: For all $b_i \in B$, as $t \to 1$, $f(b_i, t) \to OUT$.

The fact that every ball's state converges in time affords us a certain luxury. It enables us to build something extra into the formulation of the problem if desired. Suppose we added the following line: "For any ball whose state converges as time approaches the minute mark, let its state at $t = 1$ be the limit state to which it converges." Let us call this statement the *convergence premise*. If we were given this convergence premise in the formulation of the problem, then we *would* have sufficient information to logically deduce what was in the tube after the minute was up. Claim 2 would be correct: The tube would be empty! But without this convergence premise, the state of each ball, and hence of the tube itself, is indeterminate.

Note that we do *not* have the luxury of formulating the paradox of Thomson's Lamp with the convergence premise. The series in that context does not converge; it diverges.

Variations

The Ross-Littlewood Paradox can be formulated in other interesting ways. Now that we have built a logical framework for analyzing the tube and its balls, we are equipped to handle a few variations.

Victor Allis and Teun Koetsier conceived of a version as follows. At time zero, the tube is empty. After a half-minute, balls $1 - 9$ are inserted into the tube, and immediately afterward, a zero is scribbled after the numeral labeling ball 1, turning its number into a 10. After another quarter-minute, balls $11 - 19$ are inserted into the tube, and immediately afterward, a zero is scribbled after the numeral labeling ball 2, turning its number into a 20. The process continues. Allis and Koetsier observe that this version appears exactly the same as the original Ross-Littlewood version: After each step, each tube from each version effectively contains the same set of balls (if an element is defined by its current label). After step 1, each tube contains balls with labels 2 – 10. After step 2, each tube contains balls with labels 3 – 20. And so on. However, in Allis and Koetsier's version, no balls are ever physically removed, so there must be an infinity of balls at $t = 1$. This is paradoxical because although the tubes' contents are different at $t = 1$, they are the same every step of the way.

We treat the Allis and Koetsier formulation just as before. Most important, we must recognize that the history of states of the tube, and of each ball, do not logically imply any states for $t = 1$. We could deduce the states only if we were to add the convergence premise. It is also important to understand that although a given ball's label may change, its objective physical identity never does. Changing a ball's label from 1 to 10 does not remove the physical entity from the tube. As before, we can create a function that maps the pairing of each ball with a point in time to a state of that ball as *IN* or *OUT*. In that function, each ball's state converges to *IN* as $t \to 1$. This follows from the fact that once inserted, a ball is never removed. Hence, if we integrate the convergence premise into our formulation of the problem, the tube contains a countable infinity of balls at $t = 1$. There is no contradiction with the original Ross-Littlewood formulation, with or without the convergence premise.

One could also formulate a bare-bones version of Ross-Littlewood as follows. The tube starts with one ball inside, labeled 0. After a half-minute, you scribble out the zero and write 1. After another quarter-minute, you scribble out the 1 and write 2. After another eighth-minute, you scribble out the 2 and write 3. And so it goes, ad infinitum. It is left to the reader to apply the principles already discussed to this naked variation.

3.3 Laraudogoitia's Point Masses

This supertask was conceived by Pérez Laraudogoita in 1996. It invokes the concept of a *point mass*, which is a hypothetical point in space with zero dimension and finite mass. (In the 1964 James Bond film *Goldfinger*, the villain's henchman, Oddjob, crushes a golf ball with his bare hand. If he had crushed the ball until it were infinitely small, the golf ball would have become a point mass.)

Consider a countably infinite series of point masses at rest, each with mass *m*, arranged along a line that is 1 meter in length. The point masses

Figure 3.3
Point masses collide in an endless Newton's cradle.

occupy locations 0, $\frac{1}{2}$, $\frac{3}{4}$, $\frac{7}{8}$, and so on in fractional metric units. The first point mass is set in motion at a velocity $v = 1$ meter per second, aimed directly at the second point mass. After a half-second, the first point mass will collide with the second, imparting all of its momentum and energy. The first point mass is now at rest, while the second point mass proceeds with velocity v. After another quarter of a second, the second point mass will collide with the third, coming to rest after imparting all of its momentum and energy. The process continues ad infinitum, with each point mass eventually coming to rest at the location of its successor. You may be reminded of the guests at Hilbert's Grand Hotel (section 1.1), each bumping to his or her neighbor's room. You might also regard this scenario as a Newton's cradle with an infinite number of infinitely small balls (figure 3.3).

We must face a troubling reality: There is no final point mass in the series. After the passage of 1 second, no point mass comes shooting out of the other end as conservation of momentum and energy would require. This isolated system appears to have violated a very fundamental law of physics: Energy appears to have been destroyed. Conversely, if we were to rewind the clock—which our thought experiment would allow because Newtonian mechanics are time-reversal-invariant—we would see motion appear from stillness, without cause, as the particles would begin colliding with each other spontaneously. Energy would have been created from nothing. The first particle in the series would ultimately emerge, traveling in the opposite direction at 1 meter per second, coming from a system of components entirely at rest with respect to each other.

Claim 1

The laws of conservation of momentum and energy must be flawed. If we arrive at a contradiction, we must check our premises, and this supertask proves these fundamental premises to be incorrect. Furthermore, by studying the consequences of rewinding the clock, we learn that we must accept the notion of physical *indeterminism*: Not all events need a physical cause.

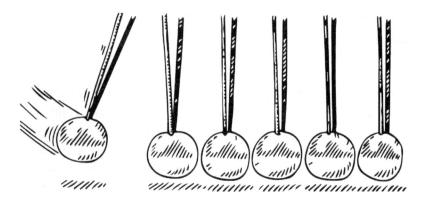

Figure 3.4
A Newton's cradle.

Claim 2

The laws of conservation and energy follow provably from a set of postulates that we have observed to be accurate in describing our world. There must be a problem with our description as to what would happen once we set the first particle in motion.

Discussion and Resolution

Claim 2 is correct. At the beginning of this chapter, we distinguished practical, physical impossibility from theoretical, logical impossibility. So far, we have untangled paradoxical supertasks with pure logic. We have not needed to say, "But that could never be done!" Our approach here is still theoretical in the sense that we are allowing a practically impossible construction. However, our construction achieves paradoxical status by invoking (and violating) certain laws of physics, so our explanation must address the physics at play. This paradox arises not from a human being performing an infinity of tasks, but rather from a human being setting a single, theoretical object in motion, just once, in the time-forward direction. Physics handles the rest. The contradiction at hand appears as a discrepancy between how we believe this isolated system would behave and certain laws of physics that we accept as absolute. To resolve the paradox, we need to address how this system would really behave on its own. We need to invoke physical principles.

Any Newton's cradle of finite length (figure 3.4) *does* conserve energy and momentum, but by creating an infinite sort of Newton's cradle, we have introduced a host of problems. It is difficult to say how the system truly would behave or in which order various catastrophes would occur. The infinity of point masses would collectively constitute infinite mass within finite volume. They would create a black hole with infinite gravitational force, causing the infinite warping and total breakdown of spacetime. The gravitational force acting between any two consecutive point masses would approach infinity as we moved down the sequence of consecutive pairs. None of the point masses actually would remain at rest; they would exert forces of arbitrarily high strength on each other, causing accelerations of unimaginably high magnitude, quickly achieving relativistic speeds with respect to each other.

There's a bandage for some of these issues, though: Recast the problem such that each successive point mass has half the mass of the previous one. Then the total mass would be finite. But this wouldn't fix a host of other problems. There would still be the problem of infinite density; each point mass would have positive mass but occupy zero volume. At the quantum level, there would be a countable infinity of pairs existing within a Planck length of each other. Classical notions of gravity and spacetime would not apply. The point masses would be floating through a putative quantum foam. There is also an important principle in quantum mechanics that limits what can be known about both the position and momentum of a particle, and this thought experiment assumes full knowledge of both—another impossibility in the construction.

These effects barely scratch the surface. The scenario simply would not play out as described in the setup.

4 Probability

Life is a school of probability.
—Walter Bagehot

If we are to understand how we form the concept of probabilities, it is useful first to understand how we form the concept of numbers. Bertrand Russell emphasized the importance of establishing a clear grammar with respect to numbers in *Introduction to Mathematical Philosophy*:

> Many philosophers, when attempting to define number, are really setting to work to define plurality, which is quite a different thing. *Number* is what is characteristic of numbers, as *man* is what is characteristic of men. A plurality is not an instance of number, but of some particular number. A trio of men, for example, is an instance of the number 3, and the number 3 is an instance of number; but the trio is not an instance of number. This point may seem elementary and scarcely worth mentioning; yet it has proved too subtle for the philosophers, with few exceptions.

Russell addresses the grammar here to distinguish the concept of a particular number from the concept of numerical quantity itself. A group of things is an element in the set of all sets having the property of a particular number; and that particular number is an element in the set of all numerical quantities. But the group of things is *not* an element in the set of all numerical quantities.

Russell uses this distinction to actually define a particular number as the set of all sets having the property of that particular number. That property is defined as the ability to create a one-to-one correspondence—a bijection—between two collections. If two collections of things can be bijected, Russell calls them *similar*. He defines the number of a class as "the class of all those classes that are similar to it," and he defines a number as "anything which is the number of some class."

Ayn Rand's definition of a concept is a "mental integration of two or more units which are isolated by a process of abstraction and united by a specific definition." In the case of particular numbers, the units are collections of things that are united by the defining property of having one-to-one relations with each other. In the case of numerical quantity, the units are the particular numbers themselves that are united by the defining property of describing some class.

If asked, very few people could offer a rigorous definition of *numerical quantity* or any particular number. Yet almost everyone forms working concepts of these at the age of three or four. Toddlers are able to use their sensory inputs to recognize numerical similarity between collections. The process of conceptualizing and abstracting is often subconscious, but that does not stop either from happening. Even if a toddler doesn't have words for sets and bijections, the toddler must rely on primitive versions of these notions to conceptualize numbers. The formation of these concepts begins with a specific kind of perceptual input: things, objects, physical entities that have shape and mass.

What are the perceptual inputs that form the basis of probability? The answer is movement, motion, action (or lack thereof)—which form the conceptual basis of *events*. Just as collections of things have the characteristic of a particular number, collections of events have the characteristic of a *probability measure*. Such a measure quantifies the relative likelihood of each event in the collection.

Many of the formal definitions in probability are complex and rely on concepts developed in measure theory. These definitions will not be required in order to understand or resolve the paradoxes presented. However, some basic conceptual building blocks will be helpful—beginning with the components of a *probability space*, which is a triplet $(\Omega, \mathcal{F}, P)$.

Here, Ω is called the *sample space*, or the set of all possible outcomes under consideration. For example, if one were to roll a regular six-sided die, the sample space would be $\Omega = \{1, 2, 3, 4, 5, 6\}$.

Further, $\mathcal{F}$ is a collection of the particular events we care about, which can be groupings of outcomes in the sample space; $\mathcal{F}$ consists entirely of subsets of Ω and has to meet certain conditions. It must contain the null set, and for every event in $\mathcal{F}$, that event's complement must also be in

$\mathcal{F}$. Stated in English, these conditions mean that we must recognize that any particular event will or will not happen; and we must also consider no outcome happening, or some outcome happening, as events. Finally, any countable union of events in $\mathcal{F}$ must also be in $\mathcal{F}$. (Meeting these conditions makes $\mathcal{F}$ a so-called *sigma-algebra*.) For example, returning to our die, we might care about the events "odd roll" and "even roll," in which case $\mathcal{F} = \{\{\emptyset\}, \{1, 3, 5\}, \{2, 4, 6\}, \{1, 2, 3, 4, 5, 6\}\}$. That is to say, the events we consider are "odd," "even," "neither odd nor even," and "either odd or even."

Finally, P is a function that maps each event to a particular probability on the unit interval. That is, $P : \mathcal{F} \to [0, 1]$. This function has to meet certain conditions, too. For instance, $P(\{\emptyset\}) = 0$ and $P(\Omega) = 1$. In English, the probability of no outcome happening must be zero, and the probability of some outcome happening must be 1. A final condition is that the probability of the union of any disjoint events must be the sum of the probabilities of those events. For example, the probability of rolling a 1 or a 2 must be the probability of rolling a 1, plus the probability of rolling a 2.

With our probability space triplet set in stone, nature makes its random choice as to which outcome $\omega \in \Omega$ occurs. Any event in $\mathcal{F}$ that contains ω is then said to have happened. Following our example, the roll of a 3 would mean that "odd" and "either odd or even" had both happened. If we were able to repeat the experiment, the relative frequencies of event occurrences would approach the probabilities assigned by P as the number of repetitions approached infinity.

Another concept that will be helpful to us, and one that is closely related to the probability space, is the *random variable*. The most formal definition of this term is too complex for our purpose here, but you can think of a random variable as a factor, condition, quantity, or quality whose value is determined randomly by nature from a collection of possible values, each with a prescribed probability measure. The outcome of our die roll, for example, is a random variable. Tomorrow's temperature at noon in your hometown is also a random variable.

If the values that a random variable can assume are finite or countable, such as the numbers on our die, the variable is said to be *discrete*. If the possible values are uncountable, like temperatures on a thermometer, the variable is said to be *continuous*.

The function that assigns probabilities to a random variable's possible values, or sets of values, is called a *probability distribution*. Probability

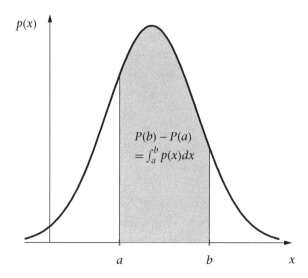

Figure 4.1
A PDF.

distributions of discrete random variables are called *probability mass functions (PMFs)*, while those of continuous random variables are called *probability density functions (PDFs;* figure 4.1). PDFs are usually denoted as $p()$.

A *cumulative distribution function (CDF)* gives the probability that a random variable will fall on or below some specified value. Its function is denoted $P()$, and it outputs a probability measure within [0, 1]. In the case of continuous probability distributions, it is calculated by taking the integral of the PDF from its lower bound up to the specified value. The probability that a continuous random variable falls within some interval [A, B] is calculated as $P(B) - P(A)$.

Whew. And here we thought defining a number was difficult. As an abstraction, probability relies on a multitude of component concepts that meet very specific conditions. Yet once again, even children are able to form working versions of probability and apply them early in life. (If I open the cookie jar, what are the chances I'll get caught? How likely is it that my parents will let me sleep over at a friend's house?)

Probability measures have some counterintuitive properties, such as this one: Any impossible event has a probability of zero, but events with a probability of zero are not necessarily impossible. Suppose your friend tells you, "I randomly picked a real number from zero to one, granting equal likelihood to each number. Try to guess the number I randomly picked." The probability of your guessing correctly is zero. Yet it's still possible for you to guess correctly. This fact follows from our requirement that P assign probabilities to measurable continua while maintaining additivity. The probability measure at a point is effectively the integral of the PDF from that point to that same point, which must equal zero; or equivalently, using CDFs, $P(A) - P(A) = 0$. By forming concepts that treat randomness methodically and self-consistently, we already have resolved what could be called a paradox.

It may also be counterintuitive that the output of a PDF can exceed 1. This fact follows from the requirement that a PDF integrate to 1 over its full domain. But how can that be so? The answer: A PDF does not output a probability measure. It outputs values that prescribe *relative* probabilities. The expression $\frac{p(A)}{p(B)}$ quantifies the relative likelihoods of A and B happening, but neither $p(A)$ nor $p(B)$ is a probability measure. Probability measures, in the context of continuous random variables, are obtained only by taking integrals over a PDF.

A natural question to ask is: What is the source of a probability distribution? We can accept probability distributions as mathematical constructs, but where do they (and their numerical prescriptions) come from? The answer depends. Treated in theory, they may be given by assumption: as quantifications of belief or as idealizations of nature. They also can govern processes in the real world. Distributions can be estimated via experimentation; data allows statisticians to form and update beliefs about true underlying distributions.

A probability distribution is inherent to any process having the characteristic of randomness. The behavior of conscious organisms, of financial markets, of mutating genes, of weather—these are only a few examples. In physics, probability distributions are crucial to our understanding of the quantum world.

The process of rationally updating probabilistic beliefs as new information becomes available is called *Bayesian inference*, and it relies on a formula called *Bayes' theorem*. Both are named after Thomas Bayes, an English statistician and philosopher, although it was Pierre-Simon Laplace who formulated the modern version of the theorem in the early 1800s. The theorem provides a formula for updating one's *prior* beliefs into updated *posterior* beliefs that are conditioned on new information. One's beliefs with respect to the likelihood of imminent rain, for example, should depend on the current state of the sky as sunny or cloudy.

Probability that takes other information into account is called *conditional* probability. It is denoted as $P(A|B)$—that is, the "the probability of A given B." The formula for conditional probability is

$$P(A|B) = \frac{P(A \cap B)}{P(B)} \tag{4.1}$$

In English, equation 4.1 reads as: "The conditional probability of A given B is equal to the probability of A and B both happening, divided by the probability of B." Using this notion of conditional probability, Bayes' theorem is stated as follows:

$$P(A|B) = \frac{P(B|A)P(A)}{P(B)} \tag{4.2}$$

Bayes' theorem can be proved quite simply. The probably of A and B must be equal to the probability of A, times the probability of B given A:

$$P(A \cap B) = P(A)P(B|A)$$

The converse is also true:

$$P(A \cap B) = P(B)P(A|B)$$

Equating the two yields equation 4.2.

No updating of beliefs is necessary if two or more random variables are *independent*—that is, if the occurrence or realization of one does not affect the probability of occurrence or probability distribution of the other. Mathematically, two random variables are said to be independent if the following holds:

$$P(A \cap B) = P(A)P(B) \tag{4.3}$$

The fact that no updating is required of independent variables can be shown by combining equations 4.1 and 4.3 to get:

$$P(A) = P(A|B)$$

$$P(B) = P(B|A)$$

The final concept required of this chapter is the expected value. *Expected value* is a probability-weighted average of all possible values that a random variable can take. It can serve as a predictor of a random variable's value, although its merits as a predictor depend on context.

Now for the paradoxes.

4.1 Sleeping Beauty

The Sleeping Beauty Paradox was originally conceived and published by Arnold Zuboff. It is the subject of extensive literature, including studies by Adam Elga, Michelle Piccione, Ariel Rubinstein, and other decision theorists, economists, mathematicians, economists, and philosophers. The *Sleeping Beauty* moniker is attributed to Robert Stalnaker. Remarkably, the problem remains a topic of ongoing debate whose resolution has not achieved academic consensus. With or without consensus, I believe that the treatment presented herein—one that is heavily reliant on analogy—is objectively true, comprehensive, and sufficient for resolving the apparent contradictions.

The story goes like this. Suppose an experimenter runs a procedure, for which Sleeping Beauty volunteers, having been fully informed about the details. The procedure begins on Sunday, when the experimenter flips a fair, two-sided coin. The experimenter never reveals the outcome of the coin toss to Sleeping Beauty. If the coin comes up heads, the experimenter puts her to sleep and wakes her the next day, on Monday, and the procedure concludes. If the coin comes up tails, the experimenter puts her to sleep and wakes her the next day, on Monday; then the experimenter gives Sleeping Beauty an amnesia-inducing drug, causing her to forget that she had ever been awakened. He then puts her to sleep (again) and wakes her the next day, Tuesday, and the procedure concludes. Upon waking this time, Sleeping Beauty is never given any new information; she is told neither the outcome of the original coin toss nor the day of the week. Therefore, the three possible waking states are indistinguishable to her. The procedure is illustrated in figure 4.2.

Now imagine that *you* are Sleeping Beauty, and you have just woken. The key question is: What is your credence that the coin came up heads? In other words, what number in [0, 1] quantifies your personal belief with respect to the probability that the coin came up heads?

Claim 1: The "Halfer" Position
The probability that the coin shows heads is $\frac{1}{2}$. We are told this is a fair fifty-fifty coin, which comes up heads or tails with equal probability. On Sunday, before going to sleep for the first time, but after the coin toss, Sleeping Beauty knows the probability of heads to be $\frac{1}{2}$. Upon any state

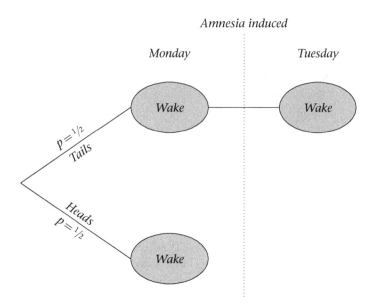

Figure 4.2
The experimenter's procedure for Sleeping Beauty.

of waking, Sleeping Beauty learns no new information about the day of the
week or the state of the coin. She has no new information that could be used
to update the probability of heads. The probability remains unconditional
and independent of her waking. It must remain as it was: $\frac{1}{2}$.

Claim 2: The "Thirder" Position
The probability that the coin shows heads is $\frac{1}{3}$. Suppose we were to repeat
the entire experiment 1,000 times. In expectation, Sleeping Beauty would
wake a total of 1,500 times; 500 of those "wakings" would follow heads, and
1,000 of them would follow tails. In fact, as the number of experimental
repetitions approaches infinity, the ratio of "head wakings" to "all wakings"
would approach $\frac{1}{3}$.

Adam Elga's argument for the thirder position works like this. Suppose
Sleeping Beauty actually were shown the outcome of the coin toss, and it
was tails. Sleeping Beauty would still have no information with respect to
the day of the week; a Monday is still indistinguishable from a Tuesday, and
so her beliefs should reflect total indifference by assigning equal credence

to both p(Monday|Tails) and p(Tuesday|Tails). Therefore:

$$p(\text{Monday and tails}) = p(\text{Tuesday and tails})$$

Now suppose instead that Sleeping Beauty were shown the true day of the week on a trustworthy calendar clock. She would still have no information with respect to the coin flip, and she could use only the fact that heads and tails outcomes are equally likely at the time of the coin flip, which might as well occur after her waking on Monday because both heads and tails outcomes are followed by a waking on Monday. Therefore, p(Heads|Monday) and p(Tails|Monday) deserve equal credence, and

$$p(\text{Monday and heads}) = p(\text{Monday and tails})$$

Through substitution, we obtain

$$p(\text{Monday and tails}) = p(\text{Tuesday and tails}) = p(\text{Monday and heads})$$

(4.4)

Equation 4.4 requires a probability of $\frac{1}{3}$ on each waking, only one of which follows heads, so the credence on heads must be $\frac{1}{3}$.

Discussion and Resolution

Before you read on, you are encouraged to spend some time working this one out on your own. Both claims seem to have sound logic, and yet they arrive at different conclusions. This paradox is profound, indeed.

The resolution proceeds by analogy. For the time being, put Sleeping Beauty in the back of your mind. (Don't forget her altogether. We'll revisit her soon.) For now, we are going to build a special candy dispenser called the Enigmatic Gumball Machine (figure 4.3), which makes its debut in this book.

The Enigmatic Gumball Machine has two *bins*, each holding a countable infinity of gumballs. Each respective bin holds red or blue gumballs exclusively. The machine is operated by three buttons: Dispense, Output, and Reset. Any time the Dispense button is pressed, the machine decides randomly whether to dispense from the red or blue bin. With probability p, it will dispense exactly 1 ball from the red bin. With probability $1 - p$, it will dispense exactly k balls from the blue bin. In this context, we will use $p = \frac{1}{2}$ and $k = 2$, to align with the Sleeping Beauty problem.

Dispensed gumballs flow into a *jar* until deliberately released to the next stage. The jar can hold a countable infinity of gumballs, so the Dispense

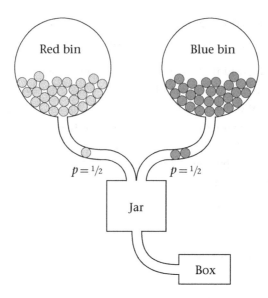

Figure 4.3
The Enigmatic Gumball Machine.

button can be pushed any number of times without risk of overflow. Each time a bin dispenses gumballs, it paints a chocolate numeral on each gumball representing that gumball's order number within the sequence of its dispensed group. Any time a ball is dispensed from the red bin, it will be labeled with the numeral 1. Any group of balls dispensed from the blue bin will be numbered 1 through k (in our case, 1 to 2). The numerals do not increase as the Dispense button is pushed repeatedly; every dispensed group will be labeled in the same manner.

If the machine's Output button is pushed, a single gumball is selected from the jar and allowed to proceed to a final *box*. The Output button chooses uniformly from among the gumballs in the jar; it grants equal probability of selection to each gumball in the jar. The box also can hold a countable infinity of gumballs without risk of overflow.

The final button is Reset, which empties the jar but not the box (see figure 4.4).

With this hypothetical gumball machine in mind, we now need a vocabulary of definitions to generalize its usage. Let us define an *action* as the single push of one of the machine's three buttons. Let us define an *operation* as a pairing between an action and some positive integer n, where

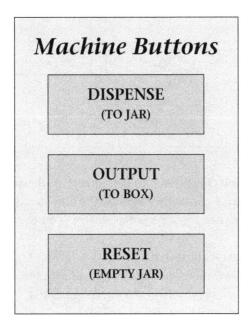

Figure 4.4
Buttons of the Enigmatic Gumball Machine.

Table 4.1
Procedure *A*.

Operation Order Number	Action	Executions
1	Reset	1
2	Dispense	1
3	Output	1
4	Reset	1

n is the number of times that action is consecutively executed (i.e., the button is pushed). A *procedure* consists of an ordered sequence of operations, and an *experiment* consists of gathering data from the repetition of a given procedure.

There are two particular procedures we care about, and we will call them Procedures A and B (see tables 4.1 and 4.2).

Though redundant, a Reset begins and terminates each procedure to ensure the purity of experimentation: The history of past procedures, or the expectation of the future of procedures, may be unknown.

Table 4.2
Procedure B.

Operation Order Number	Action	Executions
1	Reset	1
2	Dispense	$n > 1$
3	Output	1
4	Reset	1

The halfer view is consistent with Procedure A. As the number of times we perform Procedure A approaches infinity, the ratio of red to blue gumballs in the box approaches 1:1, for a probability of $\frac{1}{2}$ on each color in a single procedure.

The thirder view is consistent with Procedure B for $k = 2$. As the number of times we perform Procedure B approaches infinity, the ratio of red to blue gumballs in the box approaches 1:2 (or, in general, 1:k), for a probability of $\frac{1}{3}$ on red in a single procedure (or, in general, $\frac{1}{1+k}$).

The single Dispense action in Operation A implies uniformity of color in the jar at all times. A red and a blue gumball can never exist simultaneously in the jar if only Operation A is performed. The color that will be sent to the box is determined not by the single Output action, but by the single Dispense action, whose probability distribution is given by $p = \frac{1}{2}$. That is the only distribution that governs what gets sent to the box. This is not the case with Procedure B, which allows the jar to fill with multiple colors of gumballs at a time.

There is nothing paradoxical about attaining different results by performing different experiments with different procedures. Operation 3 differs between Procedures A and B, and therein lies the cause of divergent results.

Suppose we perform Procedure A exactly once, and we are told that the gumball in the box is numbered 1. We can now use our formula of conditional probability to update our beliefs with respect to color. The probability that a red gumball numbered 1 will be output is $\frac{1}{2} \times 1 = \frac{1}{2}$. The probability that a blue gumball numbered 1 will be output is $\frac{1}{2} \times \frac{1}{2} = \frac{1}{4}$. The ratio of the former over the sum of the two gives us the conditional probability that the output is red, given that it is labeled with a 1 and that conditional probability is $\frac{2}{3}$:

$$p(\text{Red}|1) = \frac{p(\text{Red and 1})}{p(1)} = \frac{2}{3} \qquad (4.5)$$

Equation 4.5 addresses Elga's thirder argument.

The final task is to recognize this analogy as valid. We must relate the Enigmatic Gumball Machine to our original Sleeping Beauty problem. Each construction begins with a random act of nature that sets us on a course. The gumballs themselves correspond to wakings. The numbers on the gumballs correspond to days of the week. The colors red and blue correspond to coin outcomes (heads or tails). The Enigmatic Gumball Machine, by bringing concretes (gumballs and buttons) into the construction, makes it easier to understand how divergence of results follows from divergence of procedure.

The correct answer to the problem, *as presented in this section*, is that Sleeping Beauty's credence on heads should be $\frac{1}{2}$. The problem could be stated in such a way that is analogous to Procedure B, in which case the answer would be $\frac{1}{3}$.

4.2   St. Petersburg Paradox

The St. Petersburg Paradox was conceived by Nicolas Bernoulli and appeared in a letter to Pierre Raymond de Montmort in 1713. It was named by Nicolas's cousin, Daniel Bernoulli, who offered a resolution to the paradox in 1738.

Suppose you are offered the opportunity to participate in a game that works as follows. The game begins with $2 on a table. You are given a fair, two-sided coin. You proceed to flip the coin. Each time the coin comes up heads, the amount of money on the table doubles (see figure 4.5). The first time the coin comes up tails, you take all the money on the table, and the game ends.

Assume that the offer is legitimate. Although the amount of money in the world is finite, and the rules of the game may require a payout that exceeds that amount, assume that the offer is backed by an ability to pay what is owed.

How much would you pay for the right to play this game? Think about your answer and decide on an amount before you continue reading.

Once you have chosen a value, consider the fact that the expected amount of money M you will earn by playing this game is infinite:

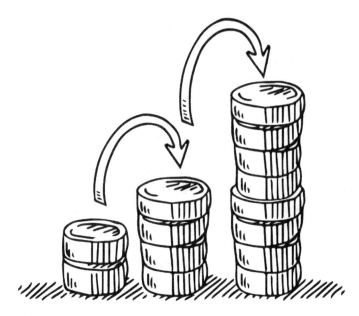

Figure 4.5
Your money doubles each time the coin turns up heads.

$$E(M) = \frac{1}{2} \times 2 + \frac{1}{4} \times 4 + \frac{1}{8} \times 8 + \frac{1}{16} \times 16 + \ldots = 1 + 1 + 1 + 1 + \ldots = \infty \quad (4.6)$$

Knowing this, do you wish to revise your willingness to pay? Would you, in fact, trade in *all* your wealth and assets for the right to play this game, whose expected payout is infinite? Or would you pay much less—even less than the amount of cash in your wallet this very moment? If the latter is true, why wouldn't you give up all your wealth in return for the expectation of infinite wealth?

Claim 1
A rational person would be willing to give up all of his or her wealth for the right to play this game because the game's expected payout is infinite.

Claim 2
It's possible for a rational person to put a small value on the right to play this game, realizing that the "expected value of the game's payout" and "personal value of playing the game" are not conceptual equivalents.

Discussion and Resolution

The central question of this paradox addresses one's willingness to pay for the right to play the game. Willingness to pay is a number, in monetary units, that reflects a subjective value judgment. The subjectivity inherent in the problem prevents us from skipping straight to a mathematical truth that resolves the apparent contradiction. However, we can form concepts that help us characterize the process by which people make value judgments with respect to currency and economic goods, defend that behavior as rational, and use mathematics to study the implications of that behavior in the context of this problem.

First, let us recognize the truthful statement in Claim 2: "Expected value of the game's payout" and "personal value of playing the game" are not conceptual equivalents. Expected value (or mean, or average) is a mathematical concept with a precise formula that defines it. Its usefulness depends on context, and the component words *expected* and *value* can be misleading. The expected value of the roll of a die, for example, is 3.5:

$$E(\text{die roll}) = \frac{1}{6} \times (1 + 2 + 3 + 4 + 5 + 6) = 3.5$$

But a die has no side with 3.5 dots; thus, it can never be "expected" to read 3.5. The expected value of this random variable is a number that not even the random variable can assume; right away, we see one possible pitfall of using expectation for the purpose of prediction. Furthermore, the word *value* refers only to the realized outcome of a random variable (a numerical quantity), not to the subjective value an individual puts on an amount of money or good. We must be careful in how we interpret expected value. Like median, mode, and quartile, expected value is a metric that helps us understand the behavior of a random variable, but we must accept limitations on its usefulness as a concept.

For this problem, we need concepts that help us describe and understand how people make value judgments with respect to money and goods. One such concept is *utility*, a degree of pleasure, satisfaction, desire, or happiness whose units are *utils*. The ownership or consumption of goods is said to provide utility to the owner or consumer. Utils are not treated as absolute units as inches or degrees are, but rather as relative units reflecting preferences over sets and quantities of goods. Economists try to infer relative utilities by observing the choices people make. People's preferences are revealed by their willingness to pay for goods. Preferences can be represented by *utility functions*, which take as inputs the amounts of various goods consumed and output some number of utils that describe the overall amount of pleasure derived. These functions can describe both the ranked order of preferences and the relative strength of preferences.

What properties would we expect utility functions to exhibit? One is the property of *diminishing marginal utility*: As a person consumes more of a good, the benefit derived from the consumption of each additional unit of that good decreases. The *marginal* utility of some good is the change in total utility with respect to that good; functionally, it is the first derivative of the utility function with respect to that good. A utility function consistent with diminishing marginal utility has a negative second derivative. We can better understand this concept through an example. If you had no shoes to wear, receiving a pair of shoes would thrill you. Receiving a second pair would also make you happy; you would not have to rely entirely on your first pair. You probably would be glad to have a third pair, but you would not value that third pair as highly as you did the first two pairs.

Because money is the medium of exchange used to purchase goods, the value of money can also be expressed with utility functions, and we would expect the law of diminishing marginal utility to apply. Once again, we can understand this by looking at an example. The value that a pauper puts on owning an extra dollar is great; it helps afford the pauper many basics of survival. A billionaire's value to owning an extra dollar is smaller; it affords little change in lifestyle.

Diminishing marginal utility implies another property of a utility function: *risk aversion*. A risk-averse person is one who tends to prefer certainty over uncertainty, even if uncertainty may yield a higher expected payoff. Insurance companies capitalize on the concept of risk aversion by charging a premium to alleviate risk. Customers of insurance are able to improve their expected utility, even if they reduce the expected value of their wealth by paying the premium. There are ways of formally defining and quantifying risk aversion, but for the purposes of this book, the high-level concept will suffice.

A popular operator among economists is the natural logarithm (figure 4.6), which is easy to differentiate and has the property of decreasing marginal returns (and hence risk aversion) in the context of a utility function. Let us choose the natural log as the basis for a utility function

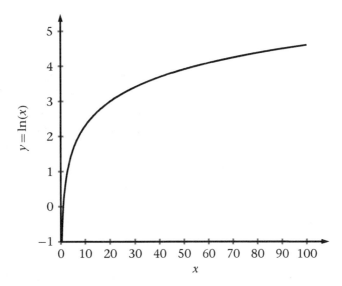

Figure 4.6
A plot of the natural logarithm.

that we will employ currently:

$$U(x) = \ln(x)$$

Here, x is some amount of money. Suppose an individual has initial wealth $w > 0$ and pays a cost c to play the game. Then that individual is indifferent between playing the game and not playing the game when the following condition holds:

$$\ln(w) = \sum_{i=1}^{\infty} \frac{1}{2^i} \ln(w - c + 2^i) \qquad (4.7)$$

The value c that solves equation 4.7 is a person's highest willingness to pay to play. It can be shown that c is finite for any $w > 0$.

The paradox is resolved, hinging on the understanding that "expected value of the game's payout" and "personal value of playing the game" are not the same. The value of money is nonlinear. Most people experience a decreasing marginal utility of money, which implies risk aversion. A rational person would be willing to pay only a finite cost to play, even if the game's expected payout is infinite.

4.3 Two Envelopes

This rich paradox has been studied by a great many scholars, including Maurice Kraitchik, J. E. Littlewood, Erwin Schrödinger, Martin Gardner, Barry Nalebuff, David Chalmers, Michael Powers, Raymond Nickerson, Ruma Falk, Ronald Christensen, Jessica Utts, Raymond Smullyan, Thomas Cover, Clifford Konold, Nelson Blachman, and John Broome. Its origins take root in similar puzzles involving neckties, purses, and packs of cards. We will focus exclusively on the context of envelopes. There are many variations in the construction of this paradox, each of which will be treated as its own puzzle with its own solution. Collectively, these solutions comprise the overall resolution to the paradox and account for total understanding. In the following discussion, the basic scenario is presented, and then numerous variations addressed.

Suppose you are asked to choose between two sealed envelopes, each of which contains money. You are told that one envelope contains twice as much money as the other. You choose an envelope at random, knowing the envelope contains an unknown amount of money x. Before opening the envelope, you are given the option of switching to the other envelope. The other envelope, you realize, contains either $\frac{x}{2}$ or $2x$. By switching, you stand either to lose $\frac{x}{2}$ or to gain x. Because you may gain more than you may lose, it appears that switching envelopes makes more sense. Yet by symmetry, there is no point in switching envelopes; if you were to switch, once again you would have the incentive to switch back. Thus, we arrive at a contradiction, and the question remains: Should you switch? More generally, under what conditions is a switch optimal, and why?

Claim 1

Yes, you increase your expected payout by switching. In fact, the expected payout from the other envelope is $\frac{1}{2}(\frac{x}{2}+2x)=\frac{5}{4}x$, as opposed to simply x.

Claim 2

There is no point in switching as the problem is stated. However, given certain beliefs, it may be worth switching—and this decision may also depend on whether you are allowed to open your envelope before deciding.

Discussion and Resolution

Claim 2 is correct. As the problem is stated, the expected gain from switching is zero, and this can be demonstrated as follows. Let the envelopes contain the values x and $2x$, for a fixed total of $3x$. This understanding is crucial: The sum of the envelopes' contents is *fixed*. With 50 percent probability, you have chosen the greater of the two and stand to lose x; and with another 50 percent probability, you have chosen the lesser of the two and stand to gain x:

$$E = \frac{1}{2}(x) + \frac{1}{2}(-x) = 0$$

Each envelope's expected value is $1.5x$.

Psychologically, however, the paradox still may seem unresolved. Suppose you are given the option of switching *after* opening your envelope. Let's say you open your envelope and find a $10 bill inside. You know the other envelope has either $5 or $20; the sum of the two is "fixed" at either $15 or $30. You still stand to gain more than you stand to lose by switching. How can we claim to have resolved the paradox?

The key to resolution is the recognition that prior beliefs with respect to the probability distribution over the contents of the envelopes are necessary to determine whether switching is optimal. As stated, the problem has provided no such beliefs. It may *feel* as if switching is optimal, but in truth, we cannot know which action to take without a prior probability distribution over the contents of the envelopes.

To form such beliefs, we must know the method by which the amounts of money are chosen and slotted into the envelopes. There is a variety of such methods, each requiring a unique treatment and implying a certain set of beliefs. Let us treat each variation as its own puzzle, with its own solution, and study how different mechanisms can give rise to different forms of optimal behavior—if and when the concept of optimality applies.

Case 1: Known Mechanism, Known Envelope

Denote the two envelopes as A and B. Consider a mechanism that fixes the value in A as a, and then chooses the value b in B as either $\frac{a}{2}$ or $2a$, with equal probability. In this case, the expected value of b is $\frac{5a}{4}$. If you are told that you are holding envelope A, then it is indeed favorable to switch, even if the value of a is unknown (i.e., envelope A is unopened).

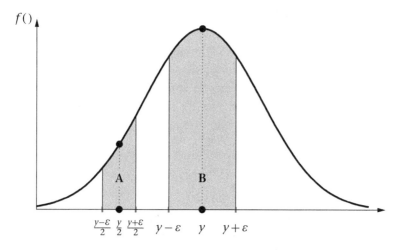

Figure 4.7
Calculating the probability that $y = x$.

Case 2: Known Mechanism and Continuous Probability Density

Consider next a mechanism that randomly chooses a value x from a probability distribution that is known to you, and then assigns the values x and $2x$ to the envelopes, which are indistinguishable to you. You open an envelope and discover a value y, which you know equals x or $2x$. We will assume a continuous probability density $f()$ over possible values of x from zero to infinity. This PDF is known to you.

The objective is to calculate a switching criterion. The probability that $y = x$ is given by

$$p(y = x) = \frac{2f(y)}{2f(y) + f(\frac{y}{2})}$$

The 2's in the numerator and denominator may seem unintuitive at first. Here's why they are there. Let's define areas A and B as shown under the curve in figure 4.7 for some tiny ε. For some y, we know the probability that $y = x$ is equal to

$$p(y = x) = \lim_{\varepsilon \to 0} \left(\frac{\int_{y-\varepsilon}^{y+\varepsilon} f(a)da}{\int_{y-\varepsilon}^{y+\varepsilon} f(a)da + \int_{\frac{y-\varepsilon}{2}}^{\frac{y+\varepsilon}{2}} f(a)da} \right) = \lim_{\varepsilon \to 0} \left(\frac{B}{B+A} \right) \qquad (4.8)$$

We can approximate the areas A and B as

$$A \approx \varepsilon f(\tfrac{y}{2}) \tag{4.9}$$

$$B \approx 2\varepsilon f(y) \tag{4.10}$$

As ε approaches zero, these expressions approach the true areas A (see equation 4.9) and B (see equation 4.10). So we can rewrite equation 4.8 as

$$p(y=x) = \lim_{\varepsilon \to 0} \left(\frac{2\varepsilon f(y)}{2\varepsilon f(y) + \varepsilon f(\tfrac{y}{2})} \right) = \frac{2f(y)}{2f(y) + f(\tfrac{y}{2})}$$

We know that $p(y = 2x) = 1 - p(y = x)$ by assumption; hence

$$p(y = 2x) = \frac{f(\tfrac{y}{2})}{2f(y) + f(\tfrac{y}{2})}$$

The criterion for switching is that *keeping what you have* must be less than the *expected value of switching*:

$$y < p(y=x) \times 2y + p(y=2x) \times \frac{y}{2} \tag{4.11}$$

Which simplifies to this rule: Switch if $f(\tfrac{y}{2}) < 4f(y)$.

Case 3: Known Mechanism and Discrete Probability Mass

Suppose the same mechanism as in Case 2: A value is randomly chosen for x from a probability distribution that is known to you. The values x and $2x$ are assigned to indistinguishable envelopes. You open an envelope and discover a value y, which you know must be equal to x or $2x$. This time, however, we will examine the case in which x takes on discrete values from a finite or countably infinite set $X = \{x_1, x_2, x_3, \ldots\}$, with the respective probabilities summing to 1, as given by a probability mass function $f()$. Let's design this function to output zero for any input that is not an element of X.

Suppose the value of y that you discover is equal to some x_i. We must compare the relative probabilities of x_i and $\tfrac{x_i}{2}$. The probability that $y = x$ is given by a more intuitive conditional probability than we saw in the continuous case:

$$p(y=x) = \frac{f(y)}{f(y) + f(\tfrac{y}{2})}$$

$$p(y=2x) = \frac{f(\tfrac{y}{2})}{f(y) + f(\tfrac{y}{2})}$$

You know immediately that if $\frac{x_i}{2}$ is not contained in X, you have opened the lesser of the two envelopes and would benefit by switching. If $\frac{x_i}{2} \in X$, we derive our switching criterion from equation 4.11 to obtain: Switch if $f(\frac{y}{2}) < 2f(y)$.

Case 4: Bayesian Interpretation with Indifference

Pierre-Simon Laplace suggested a rule for quantifying personal beliefs when states are indistinguishable. That rule is called the *principle of indifference*, and it assigns equal probability $\frac{1}{n}$ to each of a possible n states because there is no reason or information by which to weight any single event more or less heavily than the others.

Let's revisit the construction that prohibits opening an envelope before deciding whether to switch. Suppose you are told nothing about how the quantities within the envelopes are generated and assigned. All you know is that your envelope contains an amount x and one envelope contains twice what the other contains.

Applying Laplace's principle of indifference, you would realize that the two envelopes contain the amounts $\frac{x}{2}$ and x, or x and $2x$, with equal probability. You also would have no reason to believe one x to be more likely than any other. Your prior distribution for x would be uniform over all the positive reals. But suddenly the principle of indifference seems irrational because there is no uniform probability distribution over all the positive reals that satisfies the axioms of probability—namely, both countable additivity and $P(\Omega) = 1$.

Suppose, then, that you assumed or believed that the only values that could exist in the envelopes were whole-number powers of 2: 1, 2, 4, 8, 16, and so on. The principle of indifference would lead to a uniform distribution over a countable set of possibilities. Once again, this would violate the axioms of probability; such a prior belief would not be reasonable.

Any system of beliefs that puts equal probability on envelope B containing $2x$ and $\frac{x}{2}$ for any x contained in envelope A must arise from an *improper* probability distribution that violates the axioms of probability. In a sense, this statement in itself resolves much of the paradox.

What if we assumed a uniform distribution over a finite set of whole-number powers of 2? Let the set of possible values in an envelope be $X = \{1, 2, 4, 8, \ldots, 2^n\}$. Our assumption of uniformity implies that any pair $\{x, 2x\}$ is equally likely and has probability $\frac{1}{n}$, and the probability that you are holding the lesser of the two is $\frac{1}{2}$. Our beliefs no longer violate the axioms of probability, and we can calculate the expected value of switching:

$$E(\text{switch}) = \frac{1}{n} \times \frac{1}{2}(1) + \frac{1}{n} \times \frac{1}{2}(-1) + \frac{1}{n} \times \frac{1}{2}(2) + \frac{1}{n} \times \frac{1}{2}(-2)... = 0$$

If you were allowed to look inside your envelope before deciding, you would choose to switch after seeing any value except 2^n. Without the ability to peek, however, there would be no gain from switching; the expected value of switching would be zero.

Case 5: A Proper Distribution and Infinite Expectation

If you've made it this far through the Two Envelopes Paradox, then you're sure to enjoy this fascinating variant. Recall from Case 4 that any system of beliefs that puts equal probability on envelope B containing $2x$ and $\frac{x}{2}$ for any x contained in envelope A must arise from an improper probability distribution that violates the axioms of probability. It is, however, possible to construct proper probability distributions such that for any possible given value of x in an indistinguishable envelope A, the expected value in envelope B will be greater than x.

Let X be our random variable, with X and $2X$ assigned randomly to envelopes A and B. Our goal is to construct a probability distribution such that $E(B|A=a) > a$ for all a, which would be paradoxical indeed. Consider the following distribution:

$$p(X = 2^n) = \frac{2^n}{3^{n+1}} \text{ for } n = 0, 1, 2, ..., \infty$$

This is a proper probability distribution over the powers of 2 because

$$\sum_{n=0}^{\infty} \frac{2^n}{3^{n+1}} = 1$$

Suppose you choose an envelope (let it be envelope A) and imagine it contains $2^0 = 1$. Then the other envelope must contain $2^1 = 2$, so switching is optimal. But what if you imagine that your envelope A contains an amount 2^n for some $n > 1$? Then envelope B either contains 2^{n-1} or 2^{n+1}. Recall the switching criterion derived in Case 3 for discrete sets: Switch if $f(\frac{y}{2}) < 2f(y)$. Here, this switching criterion will always be met because

$$f(\frac{y}{2}) = p(\frac{2^n}{2}) = p(2^{n-1}) = \frac{2^{n-1}}{3^n}$$

$$2f(y) = 2p(2^n) = 2\frac{2^n}{3^{n+1}} = \frac{2^{n+1}}{3^{n+1}}$$

$$\frac{f(\frac{y}{2})}{2f(y)} = \frac{3}{4} \implies f(\frac{y}{2}) < 2f(y)$$

This implies that the expected value in envelope B is greater than the expected value in envelope A for *any* value in A you could imagine! We have achieved our goal: $E(B|A=a) > a$ for all a. In fact, for any $n > 1$, it can be shown that the expected value in envelope B is always 10 percent greater than a for any given a. Therefore, we don't even need to look inside envelope A to know that we want to switch to B.

How can this be? This seems crazy. From an entirely proper probability distribution, we have achieved a paradoxical result: We *always* want to switch envelopes, even if we don't know the value inside either. There must be something wrong.

In fact, there is. It turns out the expected amount of money inside either envelope is infinite: $E(X) = \infty$. More generally, any probability distribution implying $E(B|A=a) > a$ for all a will also imply that $E(X) = \infty$.

To understand why, let's revisit the switching criterion derived in Case 3. To ensure that $E(B|A=a) > a$ for all a, we must ensure that $f(\frac{y}{2}) < 2f(y)$. That is, we must ensure that each successive power of 2 has at least twice the probability of the one before it. But if that's the case, then the expected value of X will be at *least* $1 + 1 + 1 + \ldots$, meaning $E(X) = \infty$. To ensure a finite expected value of X, we must impose the condition that each successive probability is less than half of the last. The expected value will then converge, but then our paradoxical statement "$E(B|A=a) > a$ for all a" will no longer be true; instead, the overall expected value of switching becomes zero.

In the case where $E(X) = \infty$, which implies the expected values within each envelope A and B is infinite, expected value is no longer a reasonable criterion by which to judge optimal behavior. Switching envelopes without peeking yields the expected value of infinity minus infinity—an undefined quantity. We must invoke new criteria. Concepts of utility and diminishing marginal returns could be useful here, as they were in the St. Petersburg Paradox.

Case 6: Choose the Greater Number

The final variant in this book has a slightly different setup. We still have two envelopes, A and B, but rather than money, each envelope contains a positive real number. The numbers are selected by a process that could be random, but that process is entirely unknown to you. If the numbers are drawn from a joint probability distribution, you have no idea what that distribution is, or even *that* they are drawn from a joint distribution. All you are told is that each envelope contains a number and that the numbers are different.

Your task is to choose the envelope with the greater of the two numbers. You are allowed to open one envelope and then somehow decide whether to keep or switch. Believe it or not, you can actually employ a mechanism that helps you choose the greater number with a probability strictly greater than $\frac{1}{2}$.

Here's how you can achieve the seemingly impossible. First, some notation. Let us denote the values in envelopes A and B as a and b, respectively, with $a < b$, although the envelopes are indistinguishable to you. Denote the value you observe in your envelope as x, which is either a or b.

Now for the process: You must create your own random variable, Y, whose value can be any positive real number and whose PDF integrates to 1. Let the realized value of Y be y. After observing the value x in the envelope that you open, you use the following rule:

$$\text{Decision} = \begin{cases} \text{Keep} & y \le x \\ \text{Switch} & y > x \end{cases}$$

When you generate an outcome y for your random variable, you are in one of three states: either $y < a$, $y \in [a, b]$, or $y > b$. Although you cannot be sure exactly which state you are in, your decision rule can never hurt you (in a probabilistic sense), and sometimes it can help you.

In the first state, with $y < a$, your decision rule will imply *keep*, and you will end up with the greater of the two numbers with probability $\frac{1}{2}$. In this range, the strategy has given you no advantage or disadvantage.

In the second state, with $y \in [a, b]$, your decision rule will imply *switch* if you are holding the lesser of the two numbers, and it will imply *keep* if you are holding the greater of the two. In this range, the strategy guarantees you the greater envelope.

In the third state, with $y > b$, your decision rule will imply *switch*, and you will end up with the greater of the two numbers with probability $\frac{1}{2}$. In this range, the strategy has given you no advantage or disadvantage.

The probability that you will walk away with the greater envelope is, in total:

$$P(\text{Win}) = P(Y \in [a, b]) + \frac{1}{2}(1 - P(Y \in [a, b])) = \frac{1}{2} + P(Y \in [a, b])$$

This equation relies on the fact that $P(Y = x) = 0$, a characteristic of continuous PDFs. By construction, the PDF for Y is continuous over the positive reals and assigns a positive probability measure over every interval of positive length in its domain. Thus, the term $P(Y \in [a, b])$ must be positive, and the overall chance of winning is greater than 50 percent!

Any distribution for Y meeting these basic criteria will work. One good choice is a basic version of the exponential distribution:

$$P(Y = y) = e^{-y}$$

which helps us because

$$\int_0^\infty e^{-y} dy = 1$$

and

$$\int_a^b e^{-y} dy > 0$$

for any positive a and b with $a < b$. Such a basic probability distribution is shown in figure 4.8.

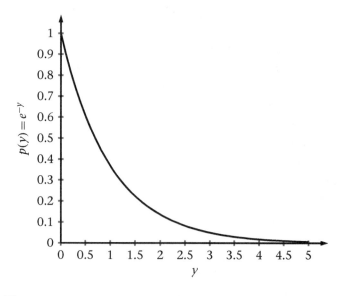

Figure 4.8
A basic exponential distribution.

Miraculous as it may seem, this mechanism gives you better than 50-50 odds of winning the greater envelope, even if you are utterly clueless as to how the values inside the envelopes were chosen and slotted. Now might be a good time to create a random number generator and start betting with friends.

4.4   Monty Hall Problem

The Monty Hall Problem, named after the original host of the game show *Let's Make a Deal*, was described and solved by Steve Selvin in 1975 in letters to *American Statistician*. The problem is mathematically equivalent to an earlier puzzle described by Martin Gardner in a 1959 column for *Scientific American*.

Suppose you are invited to be on a game show. The rules of the game are explained in advance. You will be shown three closed doors. Behind two of the doors, there's a goat, and behind the third, a brand-new car. You begin by choosing one of the doors. The game-show host, knowing which door hides the car, will proceed to open a door from the remaining two (a door you did *not* select) that he knows has a goat behind it. If both of the remaining doors contain goats, the host will randomize between the two.

Now, it's the day of the show, and you are on the air! You have chosen an initial door. The host has opened one of the other doors, showing you a goat. Two doors remain: your initial choice and one other. The host now says: "It's time to commit to a final choice. Will you keep what's behind your initial door, or do you want to switch?"

Claim 1
Each door hides the car with equal probability. You have a 50-50 chance of getting it right, so it doesn't matter if you keep or switch.

Claim 2
Counterintuitive though it may seem, you should switch doors. If you switch, the probability of getting the car is $\frac{2}{3}$.

Discussion and Resolution
Claim 2 is correct. Most people find this result difficult to believe at first. In fact, when surveyed, fewer than 20 percent of people prefer to switch. When

the columnist Marilyn vos Savant published an article in *Parade* magazine in the early 1990s defending the "switch" strategy, she received thousands of letters of disagreement, criticism, and reproach, including many from readers with PhDs in related fields. Some notable mathematicians were known to have disagreed even after seeing the proof. There is something about this puzzle that many find hard to accept.

Nonetheless, switching is definitely preferable, as the rules of conditional probability tell us. We don't even need to invoke any formulas of conditional probability to resolve the cognitive dissonance.

Let's call the door that you initially select *Door One*. When you point to Door One, one of two things is true: Either Door One hides the car or it doesn't.

If Door One hides the car, you lose by switching. If Door One does not hide the car, then by switching, you guarantee yourself a win because the remaining door that hides a goat is necessarily eliminated by the host. So really, the probability that you will land on the car by switching is equal to the probability that Door One does *not* have the car behind it. That probability is $\frac{2}{3}$.

By deliberately choosing which door to eliminate according to a specific rule—and a rule that guarantees the elimination of a goat—the host provides you with extra information that you can use to improve your odds of winning the car.

4.5   Bertrand's Boxes

The French mathematician and economist Joseph Bertrand introduced the following paradox in 1889. Suppose you have three indistinguishable boxes: a box containing two gold coins, a box containing two silver coins, and a box containing one gold coin and one silver coin. After choosing a box at random, you reach inside and remove a gold coin. What is the probability that the remaining coin in the box is also gold?

Claim 1

The probability must be $\frac{1}{2}$. We know the box is *not* the box containing pure silver. It is either the box with pure gold, or mixed gold and silver, with equal probability. Therefore, the probability that the remaining coin is gold is 50 percent.

Claim 2
The probability is actually $\frac{2}{3}$.

Discussion and Resolution
Claim 2 is correct. Suppose that instead, the question were phrased in general terms, as follows: "After reaching into the box, you take out a coin of a certain metal. What is the probability that the remaining coin is of the same metal?" Because two of the three boxes are pure (contain only one metal), two of the three boxes will yield a second coin of the same metal as the first coin. The answer to our revised question is $\frac{2}{3}$. Knowing that "certain metal" is gold does not change the underlying probability that the second coin is of the same color.

The flaw in Claim 1 lies in the assumption that our removed coin comes from the pure gold box or the mixed box with equal probability. Consider how much easier it is to draw a gold coin from a box of pure gold than it is to draw a gold coin from a mixed box. The probability of drawing a gold coin from a box of pure gold is 1, and from the mixed box, $\frac{1}{2}$. The conditional probability of having drawn from the box of pure gold, given that you have already drawn a gold coin, is $\frac{1}{1+\frac{1}{2}} = \frac{2}{3}$.

Alternatively, think of it this way: Each of the three gold coins is equally likely to be drawn from the beginning, but in two of the three cases, it will be drawn from the box of pure gold.

4.6   Two Children

Martin Gardner introduced this paradox in 1959 in *Scientific American*. The paradox centers around two separate questions. In each question, you may assume that a child is born male or female with equal probability, and that genders of siblings are independent of each other.

Question A: A family has two children. The elder child is a boy. What is the probability that both children are boys?

Question B: A family has two children. At least one of the children is a boy. What is the probability that both children are boys?

Claim 1
The answer to each question is $\frac{1}{2}$.

Claim 2

The answer to Question A is $\frac{1}{2}$. The answer to Question B is ambiguous; it could be either $\frac{1}{2}$ or $\frac{1}{3}$, depending on how the information was obtained.

Discussion and Resolution

Claim 2 is correct. Let's begin with Question A. Going forward, we'll denote the four possible gender combinations of two siblings by the elements in the set $F = \{bb, gb, bg, gg\}$. For the purposes of Question A, let the first letter in each pair represent the gender of the younger sibling. The unique information we are given—"the elder child is a boy"—reduces the possible outcomes to the set $\{gb, bb\}$, each of which occurs with equal probability. Therefore, the probability of bb is $\frac{1}{2}$.

Question B is more interesting. There are different unique processes of learning information that could ultimately lead to the same statement, "At least one of the children is a boy."

One such process would work as follows. A family is selected at random from all the families having two children. The family randomly chooses one of its children (with equal probability between the two, regardless of gender), and truthfully tells you the gender of that child. We will call this the *all-family* assumption because every family with two children is considered.

A different process could work like this. Imagine that a family is selected at random from all families who have two children, at least one of whom is male. We will call this the *limited-family* assumption because the pair of siblings was chosen from a more limited pool of families.

The all-family approach yields an answer of $\frac{1}{2}$ to Question B. Why? Initially, an element is chosen from the entire set of combinations S, with equal probability of $\frac{1}{4}$. One of the two letters of that element is then chosen uniformly, and it is shown to be b. The possible elements are now $\{bb, gb, bg\}$, with respective conditional probabilities $\{\frac{1}{2}, \frac{1}{4}, \frac{1}{4}\}$.

The limited-family approach yields an answer of $\frac{1}{3}$ to Question B because equal probability is assigned to each element of the set $\{bb, gb, bg\}$.

The resolution to this paradox lies in the understanding that both the all-family assumption and the limited-family assumption allow us to declare the same statement truthfully: "At least one of the children is a boy." However, the grammar here is not clear. That statement alone does not tell us which of the two possible assumptions was intended. So long as the intended assumption is made clear and we understand how the knowledge of that statement was produced, the problem becomes simple.

The limited-family approach can lead to a strange result if we are given more information about the siblings. Suppose that instead of selecting from all families who have "two children, at least one of whom is a boy," we select from all families who have "two children, at least one of whom is a boy with characteristic X." Let characteristic X occur independently and with probability p. That characteristic could be anything. It could mean being born on Sunday, with $p = \frac{1}{7}$; or being born during summer, with $p = \frac{1}{4}$; or having blue eyes, with $p = .08$. It simply has to be some characteristic with a known probability of occurrence. Let b_X denote a child with characteristic X.

According to Bayes' theorem (see equation 4.2) and using the limited-family approach given the previous information, the probability that both children are male is given by

$$P(bb|b_X) = \frac{P(b_X|bb)P(bb)}{P(b_X)} \tag{4.12}$$

Let's solve the right side of this equation, term by term.

$P(b_X|bb)$ is the probability that at least one of two males will have characteristic X. That probability is given by $1 - (1-p)^2$, or 1 minus the probability that neither will have that characteristic.

$P(bb)$ is the probability of two males, which is $\frac{1}{4}$.

$P(b_X)$ is the probability that at least one sibling in a pair of siblings will be a male with characteristic X. There are four pairs to consider, each with equal probability, and those are given by the four elements in the set S. We can rewrite $P(b_X)$ as

$$P(b_X) = \frac{1}{4}(P(b_X|gg) + P(b_X|bg) + P(b_X|gb) + P(b_X|bb))$$

The first term is zero because there can't be any boys in a set of only girls. The second and third terms are each equal to p, the probability that the only boy will have characteristic X. The fourth term is $1 - (1-p)^2$, as explained previously. Substituting all these components back into equation 4.12 and simplifying, we obtain

$$P(bb|b_X) = \frac{2-p}{4-p}$$

For fun, we can swap in some characteristics whose probabilities we know or can assume with a high degree of accuracy. For example, being born on a Sunday. Using the limited-family approach with the criterion "two children, at least one of whom is a boy born on a Sunday," the probability that both are male is—plugging $p = \frac{1}{7}$ into this formula—equal to $\frac{13}{27}$, or approximately 48 percent. If the criterion is being born during the first half of the year, which occurs with $p = \frac{1}{2}$, the output is $\frac{3}{7}$, or approximately 43 percent. You might have fun trying some others.

Notice what happens as you vary the rarity of the characteristic. As the characteristic becomes rarer—that is, as p approaches zero—the output ratio approaches $\frac{1}{2}$. That's because we are approaching the case in which the characteristic is almost certainly true of only one of the children, and that case is probabilistically equivalent to the case making the standard all-family assumption (where a single child is randomly sampled from all two-child families). As the characteristic becomes more common and less distinguishing or informative—that is, as p approaches 1—the ratio approaches $\frac{1}{3}$, which is equivalent to the standard selected-family case.

Ultimately, this paradox is resolved by adding clarity to ambiguous language. The result obtained under the selected-family assumption with additional information seems strange and unintuitive, but only because the process of obtaining information in that way is itself difficult to comprehend.

4.7 Simpson's Paradox

Simpson's Paradox was described by Edward Simpson in 1951 and was alluded to by several statisticians around 1900. The paradox describes a certain phenomenon that is best presented by an example. Suppose there are two universities, A and B. Each university has only two departments: an English department and a physics department. Suppose further that University A's English department has a higher ratio of women to men than does University B's English department. Furthermore, University A's physics department has a higher ratio of women to men than does University B's physics department. The question is: Because each department at University A has a higher ratio of women to men than its counterpart at University B, can we conclude that University A as a whole has a higher ratio of women to men than does University B?

Table 4.3
Ratios of men to women by department and in each university.

	English	Physics	Total
University *A*	0/10	30/40	30/50
University *B*	10/40	10/10	20/50

Claim 1

Yes. If each department at University *A* has a greater percentage of women relative to the corresponding department at University *B*, then University *A* has a greater percentage of women as a whole.

Claim 2

No. It is still possible for University *B* to have a greater percentage of women as a whole.

Discussion and Resolution

Claim 2 is correct. In general, Simpson's Paradox is the counterintuitive result, whereby a trend appearing in multiple groups of data reverses when the groups are combined or amalgamated. Consider the data given in table 4.3, for example.

In the English departments, University *A* is 0 percent male and University *B* is 25 percent male. In the physics departments, University *A* is 75 percent male and University *B* is 100 percent male. In each department, the percentage of males is greater at University *B*, and yet University *A* has a greater overall percentage of males!

An example of this occurred in 1973, when at the University of California at Berkeley, the graduate admissions ratio of men was nearly 26 percent higher than that of women. Despite the overall admissions disparity, research by P.J. Bickel, E.A. Hammel, and J.W. O'Connell (1975) shows that more departments were biased in favor of women than in favor of men. The authors conclude that "measuring bias is harder than is usually assumed, and the evidence is sometimes contrary to expectation."

The paradoxical grouping phenomenon also can be seen frequently in other contexts, such as the scoring of games. Consider, for example, three games of squash between John and Amy. Squash is won by the first player to score 11 or more points and at least 2 more points than the other player. Suppose the scores are as shown in table 4.4.

Table 4.4
Three squash games between John and Amy.

	Game 1	Game 2	Game 3	Total
John	11	11	2	24
Amy	9	9	11	29

John may have won the majority of games, but Amy earned the majority of the points. So how do we determine who is the better player? How do we use information contained in these paradoxical groupings to predict the outcomes of future games?

More generally, the real challenge of the paradox is: How do we make decisions based on data or draw inferences from data that appear to tell different stories when partitioned or aggregated? Understanding the story behind the data helps us with this task. We want to look for any causal relationships that would give us reason to partition the data.

It can be helpful to visualize the data telling a story. Figure 4.9 shows a series of data points. The *y*-values are said to be *dependent* on the *x*-values; the *x*-values *explain* the *y*-values. If we wish to model the relationship between a dependent variable and some number of explanatory variables, we can use a tool called *regression*. Regression analysis helps us interpret what kind of effect some variables have on another. One common type of regression is *linear regression*, which tries to model the relationship between variables as a straight line.

The line cutting through the data points in figure 4.9 is such a regression line, and its slope is negative (approximately −0.5). However, if we look only at the data points represented either by circles or by squares, the trend is obviously upward and positive. Be aggregating the data, we have obtained a regression line whose slope has the opposite sign to that of either partition.

Statistical (or econometric) analysis is both a science and an art. Here, the art lies in discovering and proving with a sufficient degree of confidence that the true story told by this data set is more nuanced than the one told by the aggregation. There seems be something distinct about the squares versus the circles. The explanatory variable seems to explain them differently, albeit with a positive trend for both. The squares and the circles have their own individual stories to tell as they relate to *x*.

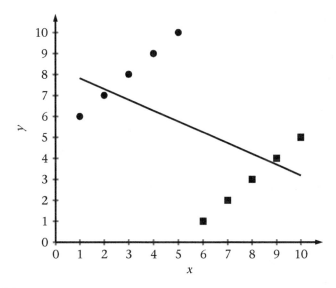

Figure 4.9
A regression with misleading results.

What were the stories told by the 1973 Berkeley graduate admissions data? Bickel, Hammel, and O'Connell (1975) conclude:

> If the data are properly pooled, taking into account the autonomy of depart-mental decision making, thus correcting for the tendency of women to apply to graduate departments that are more difficult for applicants of either sex to enter, there is a small but statistically significant bias in favor of women. The graduate departments that are easier to enter tend to be those that require more mathematics in the undergraduate preparatory curriculum. The bias in the aggregated data stems not from any pattern of discrimination on the part of admissions committees, which seem quite fair on the whole, but apparently from prior screening at earlier levels of the educational system. Women are shunted by their socialization and education toward fields of graduate study that are generally more crowded, less productive of completed degrees, and less well funded, and that frequently offer poorer professional employment prospects.

While it may appear at first glance that Berkeley on the whole discrim-inated against women, the study reveals that the overall discrimination was mildly in *favor* of women. The authors had to establish that certain departments were more difficult for either gender to enter—and that these departments attracted more female applicants. The discrimination "at a glance" was actually a result of differences of choice between genders and differences in difficulty among departments.

What kind of statistical narrative do John's and Amy's scores suggest? It is hard to say who is the better player from only three games, especially games with paradoxical scores. With more data, however—and particularly if that data included more variables—we might begin to discover an interesting story. Amy, for example, seems to be a consistent player. In each of these three games, her score was at least 9. John, on the other hand, seems to vary; he won two games by a small margin but got trounced in the third. Is there an explanation for the higher variability of John's performance? Does he have "on" and "off" days? Was he injured or tired during the third game? Had a low-calorie diet sapped his energy? Had Amy taken lessons between the second and third games? Can the inconsistency of John's game be explained by some other cause?

The *science* of statistics (and of related fields, such as econometrics and data analytics) is the development of tools—like regression models and other instruments—that are used to interpret data. The *art* of statistics is the use of those tools to discover the truth. Simpson's Paradox shows us how the misapplication of those tools can lead to false conclusions, and how their correct application can reveal truths that are contrary to intuition given certain evidence.

The simple resolution to Simpson's Paradox is the proof that paradoxical groupings, such as the ones presented here, can exist; it is to show how and why they are possible mathematically. The complete resolution tells us that discovering causal relationships is the proper way to untangle and attempt to interpret data for which paradoxical groupings exist.

5 Social Choice

Attempts to form social judgments by aggregating individual expressed preferences always lead to the possibility of paradox.

—Kenneth Arrow

For decades, the Academy of Motion Picture Arts and Sciences used a plurality vote to determine Best Picture. Over 5,000 voters would submit a single vote, picking from a few nominees, and the movie with the largest number of votes would win. In 2010, the Academy replaced plurality voting with a new system called "instant runoff voting." Instead of submitting a single choice, voters were asked to submit a preferential ranking of up to ten nominees. After all the ballots were cast, the mechanism would systematically eliminate candidates with the fewest first-place votes until only the winner remained.

Instant runoff voting offers several advantages over a plurality vote. The plurality mechanism might elect candidates that many voters strongly dislike. Plurality allows something called the *spoiler effect*—the effect of splitting votes between two similar candidates, giving an advantage to an opponent of both. Voters in plurality elections often experience incentives to misreport their true preferences and compromise, voting for a candidate with a perceived realistic chance of winning over a more personally favored choice with a perceived smaller chance. Instant runoff voting eliminates the spoiler effect and reduces the chance of electing candidates that a majority dislike.

In a 2010 article for *The New Yorker*, Hendrik Hertzberg explains the effects of instant runoff voting in the context of that year's Academy Awards:

> This scheme, instant-runoff voting, doesn't necessarily get you the movie (or the candidate) with the most committed supporters, but it does get you a

winner that a majority can at least countenance. It favors consensus. Now here's why it may also favor *The Hurt Locker*. A lot of people like *Avatar*, obviously, but a lot don't—too cold, too formulaic, too computerized, too derivative ... *Avatar* is polarizing. So is James Cameron. He may have fattened the bank accounts of a sizable bloc of Academy members—some three thousand people drew *Avatar* paychecks—but that doesn't mean that they all long to recrown him king of the world ... These factors could push *Avatar* toward the bottom of many a ranked-choice ballot.

On the other hand, few people who have seen *The Hurt Locker*—a real Iraq War story, not a sci-fi allegory—actively dislike it, and many profoundly admire it. Its underlying ethos is that war is hell, but it does not demonize the soldiers it portrays, whose job is to defuse bombs, not drop them. Even Republicans (and there are a few in Hollywood) think it's good. It will likely be the second or third preference of voters whose first choice is one of the other "small" films that have been nominated.

As a process, voting requires design. The task may seem trivial: Ask voters to express their true preferences over a set of candidates, and choose a system that fairly and justly elects a winner from the collective set of preferences. It turns out there are difficulties associated with both these steps. As you will learn from the forthcoming sections, no voting mechanism can guarantee an incentive for truthful voting with every voter, and no voting mechanism meets all supposedly reasonable criteria of justice.

A theoretical framework must be constructed to evaluate methods of aggregating individual interests into a collective decision. This theoretical framework, dating back to the work of French philosopher and mathematician Marquis de Condorcet, is called *social choice theory*.

Some structure, concepts, and terminology will be helpful in this discussion. In the context of social choice, there will be a set $V = \{v_1, v_2, ..., v_n\}$ of n voters, and a set $S = \{s_1, s_2, ..., s_m\}$ of m candidates. Each voter v_i has a preference relation p_i over the elements in S that is weakly ordered, complete, and transitive. In other words, each voter can list all the candidates by preference, those preferences are noncircular, and the voter may be indifferent toward certain candidates. The set $P = \{p_1, p_2, ..., p_n\}$ of preference relations is called a *social preference profile*. A voting mechanism, or *social choice function*, is a function that maps every possible P into a single outcome s in S, or into a complete, ranked social relation over S. A *social relation* is a ranking of the candidates in a way that "society prefers"—that is, reflecting aggregated individual preferences.

Social choice functions can take many forms. Some assign points based on rank; others employ iterative algorithms; others use pairwise comparison; and still others use different mechanisms altogether. Within the framework of social choice, there are different criteria for evaluating the fairness or justness of any given voting system.

Voters often experience the incentive to vote tactically or strategically— that is, to misreport their true preferences in the hope of electing a more favorable candidate. In the United States, where plurality voting is the norm (used to elect the House of Representatives, as well as many state and local legislatures), third-party candidates suffer from the fact that their ideological supporters may consider them unlikely winners. Libertarians and Green Party supporters have the incentive to vote Republican or Democrat, knowing that a vote for the latter two parties is a vote that "counts"—that is, it may influence the outcome over two less-preferred candidates over whom the voter still has preferences.

Strategy-proofness is considered a desirable trait in a voting system. Although it is impossible to achieve for all preference profiles meeting certain basic conditions, the incentive for tactical voting can appear in different forms and with different probabilities, depending on the size of the voter base and the voting mechanism used.

There are other fairness criteria that could be considered reasonable. One is the criterion of *nondictatorship*, which requires that a voting mechanism account for the preferences of multiple voters and not simply accommodate the preferences of any single voter. *Dictatorship*, here, is not synonymous with *tyranny* or *despotism*; a *dictator* is simply a person whose ballot "dictates" the outcome of the social choice function, regardless of the preferences of other voters.

Another criterion of fairness is the principle of *unrestricted domain*, which requires that a social choice function deterministically output a single, complete social relation for every possible social preference profile.

A slightly more complex criterion of fairness is *independence of irrelevant alternatives*. This criterion requires that an election between candidates X and Y depend only on preferences between X and Y; introducing a losing candidate Z, or changing the order between X and the losing candidate Z, should not affect the outcome between X and Y. An example may be helpful here. Suppose you are choosing between desserts at a restaurant, and the waiter offers you crème brûlée or fudge brownies. You choose crème brûlée.

The waiter returns a moment later, having forgotten to mention a third option: cherries jubilee. Having heard this third option, you change your mind from crème brûlée to fudge brownies. This example violates the principle of independence of irrelevant alternatives; the introduction of the irrelevant alternative of cherries jubilee (irrelevant because it was a losing candidate) should not have affected your choice between the first two desserts.

The criterion of *monotonicity* requires that any individual promoting a certain candidate on his or her individual preference profile—while all other individual profiles are held constant—could never cause that candidate to be demoted in the social relation chosen by the social preference function. Simply stated, a voter should not be able to hurt a candidate by ranking the candidate higher on his or her ballot.

The *nonimposition* criterion requires that the social choice function be surjective: Every social relation can be achieved by some social preference profile. The *unanimity* criterion, or *Pareto efficiency* criterion, requires that if every voter prefers X to Y, the social relation chosen by the voting mechanism must also prefer X to Y.

There are other reasonable criteria of fairness, one of which will be discussed in the next section. Some of these criteria are implied—or violated—by others. Certain fairness criteria, which we might expect to be basic to a reasonable voting system, cannot be satisfied simultaneously. Cognitive dissonance arises from the desire to achieve all such criteria of fairness, despite the impossibility of achieving them. Do mathematical impossibilities challenge the moral status of institutions of which democratic elections play an integral part? Is social choice imperfect? Is true justice unattainable? Can it be rational to wish for provably unachievable perfection?

5.1   Condorcet Paradox

The French mathematician and philosopher Marquis de Condorcet identified the concept of a *Condorcet winner*: a candidate preferred by a majority of voters in a pairwise runoff against each other candidate. That is, if X is a Condorcet winner, then a majority vote between X and any single other candidate will always elect X. Condorcet also championed the idea of the *Condorcet criterion*, a property satisfied by any voting mechanism that always elects the Condorcet winner, if one exists.

However, such a winner does not always exist, as Condorcet showed with his famous paradox. Consider three voters with the following preferences over candidates A, B, and C. Here, the symbol $\succ$ means "is preferred to."

Voter 1: $A \succ B \succ C$

Voter 2: $B \succ C \succ A$

Voter 3: $C \succ A \succ B$

The society that these individuals comprise prefers A to B by majority, B to C by majority, and C to A by majority. The preferences of each individual voter are not cyclic, but the social preference, by majority, is. It is as if the society must choose between rock, paper, and scissors. Rock beats scissors, scissors beats paper, and paper beats rock. Any choice of the three is *unstable* because there will always be a more preferred option in aggregate.

There appears to be no good option.

Claim 1

If each individual's preferences are transitive, then social preferences, as determined by the majority, must also be transitive. There must be something wrong here. A revote must be called.

Claim 2

Social preferences indeed can be transitive, even if each voter is rational, with transitive preferences. There is no obvious way to choose a winner here.

Discussion and Resolution

Claim 2 is correct. Intransitive societal preferences can be implied by exclusively transitive individual preferences. In this particular case, there is no Condorcet winner. If you are experiencing any cognitive dissonance, it likely stems from the fact that no deterministic voting mechanism is particularly defensible in this election. If that seems paradoxical, then you have presupposed that in any context, some deterministic voting mechanism *should* serve as the obvious choice. Condorcet's cycle challenges that assumption and underscores the fact that intuitively reasonable notions of fairness cannot always be met in the context of social choice. A discussion of the relationship between mathematics and fairness continues in the next section.

5.2   Arrow's Impossibility Theorem

Economist and Nobel laureate Kenneth Arrow proved that it is impossible
for an election to satisfy all of a certain set of reasonable criteria of fairness. Specifically, if there are three or more distinct candidates for which
individuals submit ranked preferences, then any social choice function that
outputs a complete social relation over the candidates cannot satisfy all of
the following:

1. Unrestricted domain

2. Nondictatorship

3. Parety efficiency

4. Independence of irrelevant alternatives

Definitions of these criteria are provided in the introduction to this
chapter. Proof of Arrow's Impossibility Theorem is omitted.

In his book *The Beginning of Infinity*, physicist David Deutsch summarizes
the disconcerting nature of Arrow's result:

> It seems to follow that a group of people jointly making decisions is neces-
> sarily irrational in one way or another. It may be a dictatorship, or under
> some sort of arbitrary rule; or, if it meets all three representativeness condi-
> tions, then it must change its "mind" in a direction opposite to that in which
> criticism and persuasion have been effective. So it will make perverse choices,
> no matter how wise and benevolent the people who interpret and enforce its
> preferences may be—unless, possibly, one of them is a dictator ... So there is no
> such thing as "the will of the people." There is no way to regard "society" as a
> decision-maker with self-consistent preferences. This is hardly the conclusion
> that social-choice theory was supposed to report back to the world.

It seems that Arrow's Impossibility Theorem delivers a deadly blow to the
very core of democratic elections, representative government, and the will
of the people. How do we reconcile Arrow's discovery with any quest for justice? In what follows, we will explore the theorem's implications, especially
as they concern the relationship between rationality and justice.

Claim 1

Arrow proved that no voting system can be deemed fair. Democratic elections of any kind are inherently flawed. We can only wish that mathematics
allowed true fairness and justice, but in our universe, mathematics has ruled
them unattainable.

Claim 2

It would be irrational to want or desire mathematical impossibility. It *would* be rational to update our concept of justice in the context of social choice to be consistent with mathematical possibility.

Discussion and Resolution

Claim 2 presents the more reasonable view. If mathematics precludes the attainment of justice—a virtue, a moral imperative—then there must be something flawed about that concept of justice. David Deutsch expands upon this idea:

> Virtually all commentators have responded to [Arrow's Impossibility theorem and other] no-go theorems in a mistaken and rather revealing way: they *regret* them. ... *They wish that these theorems of pure mathematics were false.* If only mathematics permitted it, they complain, we human beings could set up a just society that makes its decisions rationally. But, faced with the impossibility of that, there is nothing left for us to do but to decide which injustices and irrationalities we like best, and to enshrine them in law ...

> But what sort of "perfection" is a *logical contradiction*? A logical contradiction is nonsense. The truth is simpler: If your conception of justice conflicts with the demands of logic or rationality, then it is unjust. If your conception of rationality conflicts with a mathematical theorem ... then your conception of rationality is irrational. To stick stubbornly to logically impossible values not only guarantees failure in the narrow sense that one can never meet them, it also forces one to reject optimism ("every evil is due to lack of knowledge"), and so deprives one of the means to make progress. Wishing for something that is logically impossible is a sign that there is something better to wish for.

If injustice, as conceived, is encoded into our universe through mathematics, then we must reexamine our concept of justice. What fact of reality gives rise to the concept of justice and makes it useful? If we can answer that question, we can form a proper concept and determine whether it applies here. If it does not, then we should find a different way to describe the dilemma Arrow identified.

A great many philosophers have attempted to define what constitutes justice. Socrates stated the concept of justice simply as the criterion that every man get what he deserves. Ayn Rand helped us go deeper by identifying the epistemic roots of justice in *Introduction to Objectivist Epistemology*:

> What fact of reality gave rise to the concept "justice"? The fact that man must draw conclusions about the things, people and events around him, i.e., must

judge and evaluate them. Is his judgment automatically right? No. What causes his judgment to be wrong? The lack of sufficient evidence, or his evasion of the evidence, or his inclusion of considerations other than the facts of the case. How, then, is he to arrive at the right judgment? By basing it exclusively on the factual evidence and by considering all the relevant evidence available. But isn't this a description of "objectivity"? Yes, "objective judgment" is one of the wider categories to which the concept "justice" belongs. What distinguishes "justice" from other instances of objective judgment? When one evaluates the nature or actions of inanimate objects, the criterion of judgment is determined by the particular purpose for which one evaluates them. But how does one determine a criterion for evaluating the character and actions of men, in view of the fact that men possess the faculty of volition? What science can provide an objective criterion of evaluation in regard to volitional matters? Ethics. Now, do I need a concept to designate the act of judging a man's character and/or actions exclusively on the basis of all the factual evidence available, and of evaluating it by means of an objective moral criterion? Yes. That concept is "justice."

Justice is the act of judging and evaluating the character of an individual according to the available facts of his or her choices and actions, by means of objective moral standards, and treating him or her accordingly. The extent to which a person or institution (such as a judicial system) puts this act into practice is the extent to which that person or institution can be considered just.

How does this concept relate to voting? You might make the general assumption that every voter in an election *deserves* to be heard by right, to submit a ballot freely, and to have a voice that can potentially influence the outcome. This makes the unrestricted domain and nondictatorship criteria particularly attractive. The other criteria, however, have less to do with justice, or what any individual voter deserves, and more to do with the consistency of the output of the social choice function with the individual preferences that comprise its input. You might say that two of Arrow's criteria pertain to justice, and two to consistency. No voting mechanism can be both just and consistent.

Arrow's theorem does not in any way preclude the ability to judge your fellow man, his character, or just deserts. Justice, properly conceived, is and always will be entirely possible. Its possibility follows from the fact of free will.

It is also important to address the moral status of democracy as a form of government. A true democracy, in its original meaning, is a government whose every decision is made by the people. Pure democracy grants unlimited power and authority to the collective, including authority over the life, property, and individual rights of any individual member. The moral arbiter of final and *only* resort is the biggest group of people, whose judgments are unrestricted. Pure democracy is morally akin to totalitarianism; individual rights have no place in it. Constitutional republics such as the American system, by contrast, are ruled not by the people, but by laws, and ideally those designed to protect individuals' right to freedom of action, and freedom from the initiation of physical force and coercion. In a moral constitutional republic, it would not matter if a plurality or majority of citizens wanted to violate the rights of any individual. Those rights would be protected under the law. The role of democratic elections in a constitutional republic pertains to smaller details, such as the the choice of leaders and personnel who have the responsibility to uphold the law.

That is all to say that in a moral constitutional republic, Arrow's theorem poses no threat to justice or moral principles. Its only so-called threat is to the consistency of the social choice function used to elect personnel to office.

Of course, the theorem applies to any ranked voting system, not simply those in a political context—but the same principle holds: Arrow's theorem poses no challenge or obstacle to justice. To see why, we have explored the epistemic roots of justice–the facts of reality that give rise to the concept.

It is tempting to regret certain unchangeable phenomena. One could wish that the universe would naturally tend toward order rather than disorder. One could wish that printing money would make us all wealthier in real terms. One could wish for the existence of a voting system meeting criteria 1–4 given previously. But no amount of wishing will change facts. There are mathematical limitations on social choice functions, which amount to complicated versions of "You can't have your cake and eat it, too." The limitations are facts of nature. It would irrational to regret them or wish they were false.

What we can do is conceive of justice as an achievable value. Separately, we can devise rational arguments in favor of some voting systems over others, depending on context and in full recognition of mathematical limitations.

5.3   Gibbard-Satterthwaite Theorem

The Gibbard-Satterthwaite theorem is a result discovered independently by Allan Gibbard and Mark Satterthwaite in the 1970s. The theorem states that any non-dictatorial ranked voting system with three or more candidates can be manipulated through strategic voting. A social preference profile always exists, such that some voter can benefit by submitting an untruthful ballot. The proof, which is intimately related to the proof of Arrow's Impossibility Theorem, is rather unwieldy and therefore omitted.

Claim 1

Any voting system that incentivizes lying must be inherently flawed. Therefore, the Gibbard-Satterthwaite theorem proves that all voting systems are flawed. Yet again, mathematics stands in the way of justice.

Claim 2

We must accept provable mathematical truths and form concepts of fairness and justice that are consistent with those truths.

Discussion and Resolution

Claim 2 is more reasonable, and the same principled argument used in discussion of Arrow's Impossibility Theorem applies here. However, it is also worth pointing out that there is practically no reason to worry about the impossibility of strategy-proofness. The result of the Gibbard-Satterthwaite theorem is derived under the assumption that voters can see each other's ballots, which is rarely true in the real world. Voters can still vote strategically given a set of beliefs with respect to the overall social preference profile, but as those beliefs become less accurate or precise, the ability to manipulate effectively diminishes accordingly. Furthermore, as the number of voters increases, the likelihood of any single voter having the ability to affect the outcome through tactical voting decreases. Tactical voting in large populations tends to require entire strategic coalitions to work.

That said, every voting mechanism is manipulable in theory. What does this mean for justice? Nothing at all: It is important not to "surrender to inevitable injustice," but rather to conceive of justice as an achievable value. What does this mean for voting design? It is important to have rational criteria by which to choose an optimal voting system. Depending on the context, including the number of voters, the probability of tactical voting may serve as one such criterion.

6 Game Theory

I use game theory to help myself understand conflict situations and opportunities.

—Thomas Schelling

Game theory is the study of strategy. It is the mathematical field concerned with the analysis of choices made by rational players in competitive, cooperative, or noncooperative contexts. In academia, it is most commonly used in economics, political science, computer science, and psychology. Its practical applications are far reaching, including business and war.

Although game theory did not become a unique and recognized field until the twentieth century, many of its principles have been understood for millennia, as indicated by battle strategies chosen by historical military leaders and ancient texts reflecting on the incentives of individual soldiers. The field became formally recognized following the work of John von Neumann and Oskar Morgenstern, which culminated in their 1944 book *Theory of Games and Economic Behavior*. In 1950, mathematician John Nash extended and generalized their foundational work, developing a concept now known as the Nash equilibrium (to be defined later in the chapter). In 1994, Nash shared the Nobel Memorial Prize in Economic Sciences with game theorists Reinhard Selten and John Harsanyi. He became the subject of the biography *A Beautiful Mind* by Sylvia Nasar, which was adapted into the 2001 film by the same name.

A *game* is a context in which at least one player attempts to maximize his or her personal payoff by choosing from certain available actions. Typically, the payoff to a player depends not only on his or her own actions, but on the actions chosen by every other player; hence, players make decisions in anticipation of the moves or responses by other players. Games can be classified according to a wide variety of types and properties.

One property has to do with cooperativity. A *cooperative* game is one in which players can form binding, externally enforceable commitments to certain actions. A *noncooperative* game is one in which such externally enforceable commitments are not possible. Traditional game theory focuses on the latter.

Another property is simultaneity. A game is *simultaneous* if players choose their actions at the same time, or if later players are not aware of actions chosen by previous players, in which case actions are effectively simultaneous (because later players cannot base their current decisions on the knowledge of previous actions).

A game is *sequential* or *dynamic* if some players move later than others and the later-moving players have some knowledge—though not necessarily perfect information—about the moves of the earlier players. (A sequential game is said to have *perfect information* if each player knows with certainty the actions undertaken previously by all other players. Checkers, for example, is a sequential game of perfect information: Players take turns, and each player may observe the state of the board throughout the game, after every single move. There is never any uncertainty as to whether a piece is here or there. A game is said to have *complete information* if each player knows what the game itself looks like—that is, she or he knows what strategies and payoffs are available to each of the other players, but not necessarily what actions the other players are choosing.

Some sequential games are said to include *Nature* as a player. This is the terminology for describing some probabilistic, external influence to the game, made by a mover that has no personal payoff or strategic interest in the game. The Nature player is ultimately a random number generator. For example, in a game of blackjack, the dealer acts as Nature. In a game of roulette, Nature determines where the ball lands on the wheel. If a move by Nature is not visible to all players—for instance, what hands are dealt at a poker table—then the game is one of imperfect information.

Actions available to players can be discrete or continuous. Checkers, for example, offers only discrete actions; the available moves are always finite. Auctions, on the other hand, offer a continuum of actions: In theory, players can bid any nonnegative, real-number value, though the smallest monetary units, like cents, place certain practical restrictions in this scenario.

Player 2

A   B

$$\begin{array}{c|c|c|}
& A & B \\
\hline
A & a,b & c,d \\
\hline
B & e,f & g,h \\
\hline
\end{array}$$

Player 1

Figure 6.1
A normal-form game.

There are many other properties of games that need not be introduced here. The motivation so far has been to demonstrate the breadth of the concept of a game. Now let's look at how some games can be represented.

Simultaneous games are depicted in what is known as *normal form*. A normal-form game consists of a matrix that shows the actions available to the players and the corresponding payoffs. In the common two-player, normal-form game, it is standard for Player 1's available actions to appear on the left side of the matrix, labeling the rows of the matrix; Player 2's available actions appear above the matrix, labeling the columns. In figure 6.1, each player has the available actions A and B. The values appearing in the cells are the payoffs to each player for the corresponding outcome of the game. In each cell, the first value is the payoff to Player 1, and the second value is the payoff to Player 2. In figure 6.1, for example, if both players choose Action A, then Player 1 receives a payoff of a and Player 2 receives a payoff of b. Because this is a simultaneous game, neither player can wait to observe the action of the other player before choosing his or her own action. The players move at the same time—or at least, neither can observe the action of the other until having chosen his or her own action.

Sequential games are depicted in what is known as *extensive form*. Extensive-form games consist of a tree showing the possible states of the game, as well as available actions, as the game proceeds over time. Each node of the tree represents a point during the game when a player chooses between some available actions. Nodes are labeled according to the player whose turn it is to choose an action. Payoffs are written at the bottom of the tree in ascending order of the players (Player 1's payoff is listed first, Player 2's second, and so on).

Figure 6.2 shows an extensive-form game with perfect information. Player 1 makes the first move by choosing Action A or B, which Player 2 observes. If Player 1 chooses A, Player 2 then chooses Action C or D; if

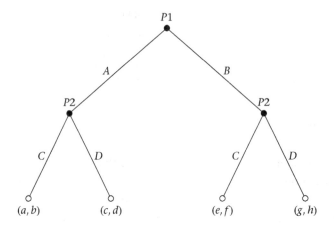

Figure 6.2
An extensive-form game.

Player 1 chooses B, Player 2 then chooses Action C or D. When it is Player 2's turn, Player 2 knows full well whether Player 1 chose A or B.

Remember that a game has imperfect information if at least one player does not have full knowledge of each other player's previous moves. In extensive form, imperfect information is represented by a dotted line connecting or encircling the nodes between which a player cannot distinguish.

Figure 6.3 displays an extensive-form game in which Player 2 has imperfect information. In this game, Player 2 moves after Player 1 but does not know Player 1's choice of action. Player 2 does not know whether Player 1 chose A or B, hence the dotted line encircling two nodes that are indistinguishable to Player 2.

The next important concept to be introduced is the information set. An *information set*—for a particular player, at a particular point in the game, and given what the player has observed—is the set of all possible moves that could have happened until now. If the game has perfect information, then each player directly observes each other player's moves: For a given stage in the game, each player knows the exact history leading to that point in the game, and so every information set in the game contains exactly one element. If, however, the game is one of imperfect information, then at least one player has at least one information set with more than one element; that player cannot be sure exactly what moves have brought the game to its current stage.

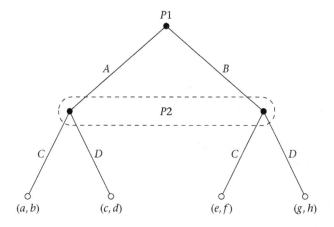

Figure 6.3
An extensive-form game with imperfect information.

In figure 6.2, which is a game of perfect information, Player 1 has one information set, and it contains a single node labeled *P*1. Player 2 has two information sets. One of them follows Action *A* by Player 1, and that set contains only the node labeled *P*2 on the left side of the tree. The other information set follows Action *B* by Player 2, and it contains only the node labeled *P*2 on the right side of the tree.

In figure 6.3, which is a game of imperfect information, Player 1 has one information set, and it contains a single node labeled *P*1. Player 2 also has only one information set, and it contains two elements, which are the nodes encircled by dashes. Player 2 cannot differentiate between those two elements when his or her information set is reached, meaning that he or she does not know whether Player 1 chose *A* or *B*.

We can now define a strategy. A *strategy* for a player is a mapping that specifies an action at each of that player's information sets. Following our examples, the strategies are as follows:

- In figure 6.1, a strategy for Player 1 is a mapping from Player 1's only information set—which is reached when the game commences—to an action *x* from among the available actions {*A*, *B*}. The same holds true for Player 2.

- In figure 6.2, a strategy for Player 1 is a mapping from his or her only information set—which is reached when the game commences—to an action *x* from among the set of available actions {*A, B*}. A strategy for Player 2 maps *each* of his or her information sets to an available action at each information set. In this case, the set of available actions is the same, {*C, D*}, at each of his or her information sets.

- In figure 6.3, a strategy for Player 1 is a mapping from Player 1's only information set—which is reached when the game commences—to an action *x* from among the set of available actions {*A, B*}. A strategy for Player 2 is a mapping from his or her only information set—the one encircled by dashes—to an action *y* from among the set of available actions {*C, D*}.

In more intuitive terms, a strategy for a given player specifies what that player does at every stage in the game for which that player legally makes a move. For a strategy to be complete, it must specify a player's actions even at stages in the game he or she doesn't think will be reached (perhaps because other players have announced their strategies, or other players' strategies are obvious). In a game with *n* players, a *strategy profile* is a collection of strategies $(s_1, s_2, ..., s_n)$—a strategy for each player.

If you created a computer program that understood the rules of some game with *n* players, and if you fed in the complete strategy profile, the computer would be able to play out the rest of the game from any given stage. For example, let's say you did this for chess, whose players are White and Black. If you fed the computer a true strategy for each player, the computer would know how to play out the rest of the game from *any* feasible arrangement of the pieces that could be reached in some game—not just from the standard initial arrangement of pieces.

We can now define the concept that forms the basis of almost all game theoretic analysis: the Nash equilibrium. A *Nash equilibrium* is a strategy profile such that, given full knowledge of all other players' strategies, no player has an incentive to change his or her own individual strategy.

The classic example used to understand Nash equilibrium is the Prisoner's Dilemma. As the story goes (illustrated in figure 6.4), Rufus and Guido have committed a crime together. The police have caught them and put them in separate interrogation rooms, so neither man knows what the other is doing. The police do not yet know the full details of the crime or the extent of each person's involvement. The police make an offer to each of the prisoners: "If you squeal on your friend, but your friend does not squeal on you, then we will be forced to believe that only your friend is guilty—in which case, we

Guido

		Cooperate	Defect
Rufus	Cooperate	3, 3	−2, 6
	Defect	6, −2	−1, −1

Figure 6.4
Prisoner's Dilemma, normal form.

will pay you handsomely to compensate you for your inconvenience, and we will jail your friend not only for his crime, but for lying to the police. If you both squeal on each other, we will know that you are both guilty, and we will jail you both for the crime, but not for lying to the police. If neither of you squeals, we will assume that you are both innocent, and we will let you split the compensation money."

There are other versions of the story. What matters is that the payoff matrix meets certain conditions of inequality. The normal form, shown in figure 6.4 has payoffs that fit this story and meet the necessary conditions. The available actions for each player in the story to choose from are to Cooperate (lie to the police) or Defect (squeal).

Can you identify the Nash equilibrium of this game?

First, you must resist the temptation to express any putative equilibrium by the associated payoffs. An equilibrium will *never* be the pair (3, 3), (−2, 6), (6, −2), or (−1, −1). Those are payoffs, not strategies. An equilibrium here is a strategy profile for two players, so it must be a collection of two strategies.

Let's consider the profile wherein Rufus and Guido both lie to the police. That profile is (Cooperate, Cooperate). Under this profile, each earns a payoff of 3. Holding Guido's strategy fixed, Rufus could improve his payoff to 6 by choosing Defect instead, and the same applies for Guido. Each has an incentive to deviate from his chosen strategy, so mutual cooperation is not a Nash equilibrium.

Now let's consider the profile wherein Rufus chooses Defect and Guido chooses Cooperate. Here, Rufus earns a payout of 6, which makes him quite happy. But Guido gets sent to jail for twice the sentence, earning a payout of −2, which he could improve to −1 by also choosing Defect. Holding Rufus's strategy as fixed, Guido could benefit by changing his strategy. (Defect, Cooperate), therefore, is not a Nash equilibrium. Neither is (Cooperate, Defect), by similar logic.

That leaves mutual defection. Under the strategy profile (Defect, Defect), neither player can personally benefit from a unilateral change in strategy while holding the other strategy constant. (Defect, Defect) is the unique Nash equilibrium of this game.

The Nash equilibrium is a mathematical concept with a rigorous definition. It is not necessarily a predictor of behavior in real-life situations. If a real Rufus and Guido were called into the police station and presented with this exact payoff matrix, there would be no promise that the observed outcome would satisfy the Nash equilibrium criterion. In real life, players may not always do their utmost to maximize personal payoff. Their personal payoffs may be influenced by other factors that are difficult to capture in a simple model—factors like empathy or a natural desire to cooperate. In real life, players may not be rational. They may not execute their strategies perfectly or have the intelligence to identify equilibrium strategies correctly. The planned equilibrium strategies of some players may not be common knowledge.

The concept of the Nash equilibrium is nonetheless helpful in analyzing situations that call for strategic thinking. Using the Nash equilibrium as a mathematical tool, we limit ourselves to contexts with rational, payoff-maximizing players who execute their strategies flawlessly, who make sound and intelligent deductions, and who believe that deviating from their own strategies will not cause other players to deviate from theirs. Players must also hold these conditions, including the one expressed in this sentence, as common knowledge.

This has been a brief introduction to some of the core concepts in game theory. What follows is a collection of game theoretical paradoxes that rely on the concepts introduced here.

6.1 Bertrand Paradox

Named after the mathematician Joseph Bertrand, the Bertrand Paradox describes a theoretical situation of competition between sellers, such that neither can earn positive profits in equilibrium.

Imagine first a market for bananas with one seller, named Sarah, and many buyers. In this market, Sarah posts a single price for all buyers to see, and any interested buyer will approach Sarah and buy one or more bananas. As a producer of bananas, Sarah faces a constant marginal cost—the amount it costs her to produce one extra banana. Suppose Sarah has information about the buyers: She knows what each of them is willing to pay. Using this information about the demand for bananas, Sarah can calculate the price at which she would earn the greatest profit. Because she is the only seller, she is a monopolist. The price she sets is called the *monopolistic price*. Sarah is quite happy in this situation, charging a price well above her marginal cost and earning a significant profit margin.

Enter Lucy, who intends to sell her bananas in the same market. Now Sarah is no longer a monopolist. She is in direct competition with Lucy, and each wants to sell bananas and maximize her own profits.

The evening Lucy arrives, both Sarah and Lucy contemplate the prices they will post independently tomorrow morning. The following facts are common knowledge between them:

- The bananas they sell are indistinguishable to the buyers.
- Their marginal costs are the same.
- The prices they post will be visible to all buyers.
- Buyers will always buy bananas from the seller with the lower price.
- If the prices are the same, each seller will capture half the market.

We will assume in this basic model that prices must be posted simultaneously and that once a price is posted, it cannot be changed.

A strategy for either player in this context is a choice of price, which we will allow to be any nonnegative real number. A Nash equilibrium is a set of strategies, one for Sarah and one for Lucy, such that neither has an incentive to deviate from her strategy given the strategy of the other.

In this model of *Bertrand competition*, wherein the sellers are competing with respect to price, the unique Nash equilibrium is the set of strategies whereby each seller sets her price equal to her marginal cost. Why is this true? Suppose Lucy and Sarah each charged a price greater than the marginal cost. If their prices were equal, each would have the incentive to drop her price by some small amount, undercutting her competitor enough to gain the other half of the market. If their prices were unequal, the higher-charging of the two would experience the incentive to slightly undercut her competitor, moving from a zero market share to a 100 percent market share. Charging a price lower than the marginal cost could never be an equilibrium strategy because in such a scenario, at least one of the sellers would lose money (or both would lose if their prices were equal), and each one could avoid that loss by adjusting her price upward. Hence, the unique Nash equilibrium occurs when Sarah and Lucy each set the price equal to her marginal cost.

The problem with setting price equal to marginal cost is that profits are driven to zero. As a monopolist, Sarah earned positive profits. As competitors, Lucy and Sarah earn nothing. This result seems paradoxical. Why would Lucy enter the market to begin with, knowing that she would earn no profits? Why would any seller enter a competitive marketplace in the real world if this is what ultimately happens?

Claim 1

The results of this model should discourage real-life entrepreneurs from entering competitive markets. Economic theory, which holds that price competitors like Sarah and Lucy earn zero profits in equilibrium, will prevail in practice.

Claim 2

This theory should be taken with a grain of salt. Competition does drive prices down in reality, but rarely to the point where the seller earns zero profit. There are a number of complicating factors in real life that prevent markets from behaving this way in practice.

Discussion and Resolution

Claim 2 is the correct view. Theoretical economic models boil down to math problems whose frameworks should reflect real life as closely as possible, while remaining tractable. Sometimes the simplifications made during the construction of the models cause the solutions to these math problems to depart from observations. There are important lessons to be learned from economic models, but economists must recognize their limitations.

In the case of Bertrand's Paradox, there are a few reasons why the equilibrium outcome of the model is not what we see in real life. The first reason is that products are rarely exactly the same. There is often some amount of differentiation, potentially as insignificant as branding, that might prevent buyers from feeling indifferent between products simply because their price is the same. Even if products are exactly the same, the sellers may be differentiated by reputation and other factors. Trust and reputation are meaningful to buyers. Buyers may be more willing to do business with companies they trust because they believe that they will save time and avoid hassle, both of which are costly. Sellers are also differentiated by their geographic proximity to buyers.

Pricing information is not freely available. It takes time to scope the market and compare prices, and time is costly to the buyer, meaning that a buyer may be willing to settle for a higher price if doing so saves shopping time.

Sellers may be constrained by capacity, unable to produce enough to satisfy the entire market by themselves. Capacity constraints would make room for competitive sellers to earn profit.

If the price-posting game is infinitely repeated, then even in theory, Nash equilibria can be supported such that prices are above the marginal cost. Sellers can agree to cooperate by setting prices that are higher than marginal cost, so long as this game continues indefinitely. Each seller can pose a credible threat to the other, forcing a mutual return to "price equals marginal cost" if either should deviate from the cooperative agreement on any given iteration of the game.

You may be able to think of other reasons why competition does not entirely drive out profits in the real world, despite the equilibrium result of the Bertrand model of price competition. Ultimately, the resolution to this paradox lies in understanding that theoretical economic models are limited in their ability to incorporate complexities of human behavior and human life.

6.2 Braess's Paradox

Formulated by the mathematician Dietrich Braess in 1968, Braess's Paradox describes a situation wherein adding a road to a traffic network actually can increase the total travel time for each car.

Suppose 1,000 cars wish to travel from a start point to an end point, as shown in figure 6.5. Initially, cars must choose between two routes: one passing through point A and one passing through point B. The travel time along a segment of road may be fixed, or it may depend on the amount of traffic on the road. In the illustration, T stands for traffic, meaning the total number of cars electing to use that road. The travel time between the starting point and point A is $\frac{T}{100}$, meaning the total number of cars using

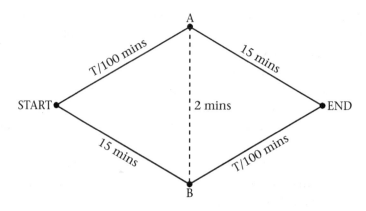

Figure 6.5
A paradoxical road network.

that segment divided by 100. The travel time between point A and the end is fixed at 15 minutes. The route passing through point B is analogous but flipped in terms of individual travel times. The characteristics of this road network are common knowledge to all drivers.

Suppose in this initial model there is no road connecting points A and B. Nash equilibrium is reached when exactly 500 cars choose the route through A and 500 choose the route through B. The travel time through each traffic-dependent road is $\frac{500}{100} = 5$ minutes. The travel time through each fixed-time road is 15 minutes. Therefore, the total travel time for any given car, in equilibrium, is 20 minutes.

To see why this is an equilibrium, consider any situation in which X cars choose the A route, and $1,000 - X$ cars choose the B route, where $X \neq 500$. Calculate the travel time along each route, and you will see that the travel times are unequal, meaning that some cars have an incentive to deviate to the shorter route. The presence of such an incentive means that equilibrium has not been reached.

Let us now add a new road to the network, connecting points A and B. Suppose this road requires a fixed travel time of 2 minutes to traverse. In equilibrium, *all* cars will now follow a route from the starting point to point A, then to point B, then to the end. The travel time from the start to point A, which is the same as the travel time from point B to the end, is $\frac{1000}{100} = 10$ minutes. All in all, this route takes 22 minutes to travel in equilibrium.

We know this is an equilibrium because no single driver can shorten his or her travel time by choosing another route. The route from the start to point B to the end would take $15 + 10 = 25$ minutes. The route from the start, to B, to A, to the end would take $15 + 2 + 15 = 32$ minutes.

Adding a road to this network, without taking any roads away, has in fact increased the equilibrium travel time. How can this be?

Claim 1

Adding a road to a traffic network could never increase the equilibrium travel time. Giving drivers more options can only reduce equilibrium travel time, just as opening a new lane on a freeway can only serve to reduce traffic. The logic here must employ some trick of arithmetic.

Claim 2

The logic here is sound; adding a road to a traffic network can indeed increase the equilibrium travel time. Conversely, removing a road from a network can reduce equilibrium travel time.

Discussion and Resolution

Claim 2 is correct. There is nothing faulty with this logic. Paradoxically, connecting points A and B increases the equilibrium travel time.

This happens because each driver makes a decision that is individually rational: in his or her own best interest, but not in the interest of reducing the total travel time across all drivers. The result is a form of the *tragedy of the commons*, a situation in which rational, utility-maximizing individuals ruin or exhaust a shared resource through action that is collectively inefficient.

Possible instances of Braess's Paradox have been observed around the world, notably in South Korea, Germany, and the United States. The paradox has analogous results in other fields of application, including electrical systems (e.g., power transmission networks) and biological and ecological systems (e.g., food chains). A theoretical argument can be made for the existence of the paradox on a basketball court, when one views a team of players as a network of routes to scoring a basket.

6.3   Parrondo's Paradox

Parrondo's Paradox, described by Spanish physicist Juan Parrondo, describes a situation in which the combination of two losing games creates a winning game. Here, a losing (winning) game is one that creates a negative (positive) payoff to the player when played repeatedly on its own. A very simple example of this paradox is presented next.

A player begins by putting some integer number of dollars into a pot. At any given time, the amount of money in the pot is said to be the player's current capital.

Consider Game A, which works as follows. The player flips a coin, and regardless of the outcome of the toss, his or her current capital decreases by $1. This is obviously a losing game. The coin toss makes it feel like a game, but the toss isn't really necessary. The player loses $1 every time this game is played, no matter what.

Now consider Game B, which works like this. If a player's current capital is an odd number, he or she loses $5. If it's an even number, he or she wins $3. For every consecutive two rounds this game is played, the player suffers a net loss of $2, for an average loss of $1 per round. Therefore, this is also a losing game as defined here.

Play either game exclusively, and you will eventually lose all your money. But what if you alternate between these two games, starting with the more profitable game given your initial capital?

Suppose you start with $10. It makes sense to begin with Game B, which earns you $3 since ten is an even number. Your current capital is now $13. Then you play Game A, losing $1, which brings your current capital to $12. Now you play Game B again, which earns you $3 since twelve is even, bringing your current capital to $15. And so on. Playing *BABABA...* nets you $2 every two rounds, for an average of $1 per round.

Alternating between the two games is definitely a winning endeavor, but either game on its own is a losing endeavor. How can this be?

Claim 1

A combination of losing endeavors cannot become a winning endeavor. There must be an arithmetic mistake in the example.

Claim 2

A combination of losing endeavors can become a winning endeavor. In the example given here, the capital-dependent nature of Game B allows this to happen.

Discussion and Resolution

Claim 2 is correct. A variety of Parrondo games can be created, with varying levels of complexity. In the example given here, the payoffs are deterministic, as is the strategy of alternation between A and B. The paradox still can manifest in constructions that have probabilistic payoffs and alternations, so long as the underlying probabilities meet certain criteria.

The characteristic that gives rise to the paradox is the state-dependent nature of at least one of the games. In this case, the parity (oddness or evenness) of the amount of capital in the pot governs the outcome of Game B. The state of parity changes each time a game is played, allowing the player to selectively take advantage of the $3 upside in Game B, while minimizing the downside. Other kinds of Parrondo games work by the same principle of state dependency.

Parrondo conceived of this paradox in connection with the Brownian Ratchet, which is described in chapter 11.

6.4   Rubinstein's Email Problem

In 1989, the economist Ariel Rubinstein introduced a game resembling what is known as the Two Generals' Problem, also called the Coordinated Attack Problem. The essence of Rubinstein's game is captured in the following original scenario. (Much creative liberty is taken with this backdrop. Rubinstein's paper had no aliens and no battles.)

Suppose you are the commander of an Earth space fleet. You have allied with the Darunian alien race to defeat your common enemy, the Zadraki, who inhabit the planet Zadrakis. You and the Darunians have surrounded Zadrakis with your spaceships and are attempting to coordinate an attack. You and the Darunians each must choose independently between two actions: fight defensively or offensively.

Meanwhile, the Zadraki are aware of the imminent attack and are choosing randomly between two strategies of their own. The Zadraki have two possible actions: They will fight either with death beams or with a photonic cannon. It is common knowledge to you and your Darunian allies that the Zadraki always use the death beams with probability $1 - p$, and the photonic cannons with a probability p, where $p < \frac{1}{2}$.

Consider the situation in which the Zadraki choose death beams. If both allied armies fight defensively, the beams will reflect off your combined shields in such a way as to defeat the enemy. If you both fight offensively, the Zadraki will retreat to their underground dwellings undefeated; you and the Darunians gain nothing and lose nothing. If you and the Darunians choose different strategies, the offensive force gets wiped out by the death beams, the Zadraki are undefeated, and the defensive ally is unharmed.

Now consider the situation in which the Zadraki choose the photonic cannon. The cannon takes a long time to power up. If both allied armies attack offensively, they will be able to destroy the cannon swiftly by their combined military forces, achieving certain victory. If both allies choose a defensive posture, they will not be able to overcome the cannon, and the battle remains at a standstill, with neither ally gaining or losing anything. If one ally fights defensively and one offensively, the offensive army—being smaller on its own—will be wiped out by the cannon, while the defensive army gains nothing and loses nothing.

Fortunately, your own military has developed a spying technology. You can see in advance, with 100 percent accuracy, whether the Zadraki have elected to use the death beams or the photonic cannon. You and the Darunians have created a messaging system to help share information. As soon as your spying machine informs you of the Zadrakis' random choice, your computer is programmed to send an automatic message to the Darunians if (and only if) the Zadraki are using the photonic cannon. Furthermore, your computer and the Darunians' computer are both programmed to send automatic confirmations of receipt of *any* message received, including confirmations of confirmations. So, if you see the Zadraki using the photonic cannon, your computer sends a ping to the Darunians, which triggers a ping in response from them, which triggers a ping in response from you, and so on. If you observe a choice of death beams by the Zadraki, your computer sends nothing.

There is one snag. Due to interference from a nearby star, there is some small probability ε that any message sent from either computer will fail to be delivered. Eventually, when your computer sends a message that receives no reply, you are not sure whether it was your computer's last message, or the Darunians' confirmation of it, that failed to transmit. As soon as this radio silence occurs, your computer tells you exactly how many pings it has sent (regardless of whether they were received). At that point, you have no further way of communicating with the Darunians. You and the Darunians then must each decide whether to play defensively or offensively.

Let's look at a few other details. Assume that losing one's army in battle implies a negative payoff of greater magnitude than that achieved when the enemy is defeated. Assume also that the nature of the game is common knowledge to both you and the Darunians. The only thing that is not common knowledge is the Zadrakis' choice of weapon. Even if you know that they know that you know that they know (and so on, for some finite number of times) which weapon the Zadraki are using, this knowledge is not "common." It would be common if and only if the chain extended infinitely.

The sure way to victory is for you and the Darunians to coordinate your attacks by playing jointly defensively when the Zadraki use the death beams, and jointly offensively when the Zadraki use the photonic cannon. Now, the big question: How many confirmations do you each need to see before such coordinated behavior can be sustained in equilibrium, ensuring your victory?

Claim 1

Coordination can be sustained in equilibrium so long as you send the Darunians an initial message, receive confirmation from them, and send a confirmation of their confirmation.

Claim 2

It cannot be calculated exactly how many confirmations are required to sustain coordinated behavior in equilibrium, but at some point, common sense kicks in. Ten confirmations should do the trick; 10,000 confirmations will almost certainly do the trick; and 1 million confirmations will absolutely do the trick.

Claim 3

There is no finite number of transmissions that will support the equilibrium of coordinated behavior required to guarantee victory. This is true even if

ε is arbitrarily small. The messaging system that you have set up with the Darunians offers no benefit.

Discussion and Resolution

Ariel Rubinstein presented a generalized form of this problem and proved the paradoxical Claim 3 to be correct. In his version of the problem, unreliable emails sent automatically between computers are the only hope for establishing coordinated behavior. Some notation will be helpful in what follows. Suppose that you are Player 1 and the Darunians are Player 2. The state of nature is the weapon w, randomly chosen by the Zadraki, which is either A (death beams) or B (photonic cannon). Let T_i denote the number of transmissions sent by Player i.

Let's proceed in small steps, working toward an understanding of Rubinstein's paradoxical result.

Information Sets and Strategies

The first step to analyzing this coordination problem is understanding: What is a strategy for you, and what is a strategy for the Darunians? Remember that a strategy is a mapping from *every* one of a player's information sets to a choice of available actions. So, we should start by understanding the information sets.

You have one information set following the state of nature $w = A$, and a countable infinity of information sets following the state of nature $w = B$. In state A, $T_1 = 0$. In state B, T_1 is a positive integer. In state B, each information set has exactly two elements, or positions of the game that are indistinguishable to you: Either your computer's last message failed to transmit or the Darunians' confirmation of it failed to transmit.

The Darunians have a countable infinity of information sets because T_2 can be any natural number. Each of these information sets has exactly two elements. If $T_2 = 0$, the Darunians do not know whether the state of nature is A or whether the state of nature is B and your initial message failed to transmit. If $T_2 > 0$, the Darunians do not know whether their last message failed to transmit or whether your confirmation of it failed to transmit.

A *strategy* for Player i, which we will denote as S_i, is a function that maps every possible $T_i \in \mathbb{N}$ to an action in the set {Defensive, Offensive}. ($\mathbb{N}$ is the symbol for the set of natural numbers.) A Nash equilibrium in this game is

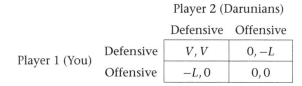

Figure 6.6
Game A: Zadraki choose death beams, probability $1 - p$.

		Player 2 (Darunians)	
		Defensive	Offensive
Player 1 (You)	Defensive	$0, 0$	$0, -L$
	Offensive	$-L, 0$	V, V

Figure 6.7
Game B: Zadraki choose the photonic cannon, probability p.

a pair of strategies (S_1, S_2), one for you and one for the Darunians, such that each strategy is payoff-maximizing given the other. We can interpret $S_i(t)$ as the action chosen by Player i when his or her computer sent t messages.

Games in Normal Form

The normal-form games are shown in Figures 6.6 and 6.7. Let V be the magnitude of the payoff to an army after victory and L be the magnitude of the payoff to an army after loss (i.e., getting wiped out). It was given in the problem that $V < L$ and $p < \frac{1}{2}$.

The normal-form games make it easier to see that playing offensively is dangerous for any player who doesn't know that the ally will join in the offense. Any player who plays offensively, while the other plays defensively, is sure to get wiped out. A defensive play is safer under this kind of uncertainty, but it never achieves victory in Game B.

Each of the normal-form games shown here, considered on its own—assuming away any uncertainty as to the state of the world, and hence assuming away the messaging system—has two Nash equilibria. Both {Defensive, Defensive} and {Offensive, Offensive} meet the condition of equilibrium, even if one equilibrium may be preferable to the players over another. A profile of strategies qualifies as an equilibrium, so long as no

Table 6.1
Probabilities of transmission pairs.

(T_1, T_2)	Probability
$(0, 0)$	$1 - p$
$(n+1, n)$	$p\varepsilon(1 - \varepsilon)^{2n}$
$(n+1, n+1)$	$p\varepsilon(1 - \varepsilon)^{2n+1}$

single player has the incentive to unilaterally deviate, taking all other strategies as given.

Probabilities

We know that the difference between T_1 and T_2 will either be zero or 1. The probability of any pair (T_1, T_2) is given in table 6.1.

Proof of Rubinstein's Paradoxical Result

Rubinstein uses mathematical induction to prove his paradoxical result. Readers who are unfamiliar with this method would benefit from reading the introduction to chapter 8 of this book. Our goal is to determine whether an equilibrium exists that can support optimal or coordinated behavior— that is, whereby both players play defensively in State A and offensively in State B.

Suppose there is a Nash equilibrium (S_1, S_2), such that $S_1(0) = $ defensive. The interpretation is that in this equilibrium, when you observe the state of the world to be A and send no messages, you play defensively.

Consider the case when $T_2 = 0$, meaning that the Darunians received no message from you, either because you never sent one and $w = A$ (with probability $1 - p$), or because it was lost and $w = B$ (with probability $p\varepsilon$). If the first case is true, from our initial assumption, the Darunians know that you play defensively. Even if the Darunians do not know what $S_1(1)$ is, they can calculate the *worst* payoff they can receive by playing defensively, and the *best* payoff they can receive by playing offensively:

$$\text{Worst payoff to choosing } S_2(0) = \text{Defensive: } \frac{(1 - p)V + p\varepsilon 0}{1 - p + p\varepsilon}$$

$$\text{Best payoff to choosing } S_2(0) = \text{Offensive: } \frac{(1 - p)(-L) + p\varepsilon V}{1 - p + p\varepsilon}$$

The worst defensive payoff is better than the best offensive payoff, so it is strictly optimal for the Darunians to play defensively when $T_2 = 0$. We now know that $S_1(0) = S_2(0) =$ Defensive and have completed the *base step* of our proof by induction.

The next step is the *inductive* step. Let us make the inductive assumption that for all $T_i < t$ for some t, both you and the Darunians always play defensively in equilibrium. Assume that $T_1 = t$. In this situation, you do not know whether $T_2 = t$ (the Darunians received your tth message, but their confirmation was lost), or $T_2 = t - 1$ (the Darunians never received your tth message). From your point of view, the conditional probability of the second scenario (denoted z) is

$$p(T_2 = t - 1 | T_1 = t) = z = \frac{\varepsilon}{\varepsilon + \varepsilon(1 - \varepsilon)} > \frac{1}{2}$$

The inequality implies that the second scenario is more likely than the first; it is more likely that the Darunians never received your message. Our inductive assumption tells us that $S_2(t - 1) =$ Defensive. Even if you do not know what $S_2(t)$ is, you know that the Darunians are more likely to play defensively. You know that the state of the world is B, meaning that if you play defensively, your payoff will be 0. If you play offensively, your expected payoff is at most

Best payoff to choosing $S_1(t) =$ Offensive: $z(-L) + (1 - z)V$

Because $z > \frac{1}{2}$ and $V < L$, it is optimal for you to play defensively, hence $S_1(t) =$ Defensive. Understanding this logic, the Darunians similarly choose $S_2(t) =$ Defensive.

Together, the base step and inductive step are sufficient to show that if you play defensively in State A (as the optimal, coordinated equilibrium of that individual game requires), then both you and the Darunians will always play defensively in State B as well. This is true regardless of how little noise exists in the transmission network (i.e., how small ε is) and how many transmissions have been sent. Even if both allies know that the state is B, and each knows that the other knows that the state is B, and so on—thanks to any arbitrarily large number of transmissions, even trillions upon trillions—equilibrium can never support a coordinated offense.

What Would You Do?

Rubinstein's proof highlights the difference in strategic behavior under "common knowledge" and "almost common knowledge" scenarios, the latter meaning when the number of messages sent is very large. The fact that equilibrium cannot support coordinated behavior under almost common knowledge seems to violate intuition. If your computer had sent fifty messages to the Darunians, what would you do? Would you play offensively or defensively? What if your computer had sent 10 billion messages? Would you be willing to choose a nonequilibrium strategy and count on the Darunians to do the same?

7 Self-Reference

I wish my wish would not be granted!
—Douglas R. Hofstadter

Answer truthfully, yes or no: Is "no" the correct answer to this question?

Think about it a moment. Whether you answer "yes" or "no," you invalidate your answer. You cannot return a correct response to that question. Whatever you say will be made false by the very fact of your saying it.

Now answer this one truthfully: Is "yes" the correct answer to this question? This time, your answer validates itself either way. Whatever you say will be made true by the very fact of your saying it.

What about, "Does this question have an answer?" If you answer "no," you have given the question an answer, but denied the existence of one—a contradiction. If you answer "yes," you have given the question an answer, and confirmed that it has one—a consistency. It would appear, then, that this rather abstract question does have a single correct answer.

These brain teasers hint at the kind of intrigue that follows in this chapter, and they all have something in common: They involve self-reference.

Self-reference can appear in propositions, in definitions of objects, and elsewhere. Some of the deepest, hardest, and most ancient logical and philosophical problems fall into the family of those caused by self-reference, but not every instance of self-reference drags us into the lion's den of paradox. We use self-reference in natural languages frequently, every time we use the word *I*. Why do some instances of self-reference cause problems, but not others? That is a difficult question, and one that this chapter will help answer.

You may recall from the introduction to this book that Ayn Rand defines logic as the "art of non-contradictory identification." The history of philosophy points to a single father of this art: Aristotle. In a series of works collectively entitled the *Organon*, Aristotle lays out the oldest foundation of logic, introducing several key concepts. Among these is the concept of *premises*, or things supposed, which necessitate a new conclusion not originally stated as a premise. The process of moving from premises to a new conclusion is called *deduction*. The other form of argumentation Aristotle introduces is that of *induction*, which is discussed in chapter 8. More than 2,000 years later, many of Aristotle's foundational concepts of logic are still used, although great advances have been made in formalizing the art, expanding its scope, and overcoming its limitations.

Today, logic is embedded in the tapestry of formal systems and languages. A *formal system* is a system of abstract thought that serves as the foundation for deriving logical and mathematical concepts. There are four basic components to a formal system. First, there must be a finite alphabet of *symbols* that can be arranged into strings, creating formulas or statements. Second, there must be a *grammar* that specifies how a well-formed formula is constructed from the alphabet of symbols. Third, there must be a set of *axioms*, the most primitive formulas or statements taken for granted, which are themselves well formed. Finally, there must be a set of *inference rules* that tell us how deductions can be made.

Semantics is the study of meaning. It examines the relationship between symbols, words, and phrases and what they mean. The grammar of a formal system, combined with the meaning attached to grammatical expressions, together comprise a formal language. *Formal languages* are created with a specific purpose—for example, to support a foundation for all of mathematics, or to write computer programs—while natural spoken languages have evolved over millennia to help humans communicate. Natural languages can have subtlety, nuance, ambiguity, double entendre, beauty, and artfulness of expression and interpretation. Natural languages evolved to reflect and accommodate the emotional complexities we experience as human beings. Formal languages, by contrast, are designed specifically for certain applications, like denoting relationships between numbers and other mathematical objects, or expressing computations.

Within a language, certain ideas can be expressed in the form of propositions. A *proposition* is a meaningful, declarative statement that is either true or false. A proposition's truth or falsity is called its *truth value*. A proposition consists of a subject and a predicate. The *subject* is what the proposition is about; the *predicate* contains a verb and expresses some idea about the subject. A proposition is a semantic notion. It carries meaning. It is possible for different sentences to have the same meaning and therefore express the same proposition. A single sentence can contain multiple propositions, and it is also possible for sentences to express but not assert certain propositions. Let's look at some examples:

- *Please come here.* This is a sentence, but not a proposition, because it is an imperative giving a command, not a declarative with a truth value.
- *Mount Everest has snow.* This sentence asserts a true proposition.
- *Mount Everest has no snow.* This sentence asserts a false proposition.
- *I haven't eaten in five hours, and I am hungry.* This sentence asserts two propositions.
- *The judge ruled that Jack committed a crime.* The entire sentence itself is the asserted proposition. *Jack committed a crime* is also a proposition, which is expressed but not asserted.

If we are to claim a proposition is true or false, we had better have a good definition of truth and a standard for evaluating it. As it turns out, truth is not so easy to define formally. Aristotle defines *truth* and *falsity* as follows:

> To say of what is that it is not, or of what is not that it is, is false, while to say of what is that it is, and of what is not that it is not, is true.

More simply put, according to Aristotle, truthful propositions call things as they are, and false propositions call things as they are not. The mathematician and logician Alfred Tarski offers a similar definition:

> A true sentence is one which says that the state of affairs is so and so, and the state of affairs indeed is so and so.

These definitions seem desirable in their simplicity and adequate in their meaning. But as you will learn from this chapter, using either of these concepts of truth, it is not always easy to evaluate certain declarative statements as true or false. This chapter will demonstrate the need for an unambiguous and comprehensive definition of truth, and a rigorous analytic toolkit for its study. The study of truth here does not mean the pursuit of all knowledge in all fields, but rather the study of the *concept* of truth itself, in a formal theory.

The paradoxes of self-reference tend to be semantic, set-theoretic, or epistemic (pertaining to knowledge and beliefs) in nature, and many have similar structures. Paradox in this chapter generally arises when either (1) members of mathematical collections are defined by means of reference to the entire collection, or (2) propositions speak to their own truth values or the implications of their own truth values. The resolutions tend to argue that these definitions are actually meaningless and conceptually invalid—and then they proceed by actually changing the formal system or language from which the paradox emerged. By analogy, the paradox is an instability in a skyscraper, which is fixed by altering the construction of the skyscraper's foundation, down to the bedrock.

This is drastic, and it poses interesting philosophical questions. It seems reasonable that languages, both formal and natural, should be allowed to express statements that are false or absurd. We don't change the English language to preclude us from expressing absurdities like *Lions are tigers*, nor do we change the language of mathematics to preclude the expression of contradictions like $1 = 0$. Why, then, should some statements or mathematical objects—respectively grammatical and definable—be so problematic as to compel us to change the systems used to express them? The problem occurs when some valid inference within the system actually supports a contradiction. If one used valid rules of inference to conclude that *Lions are tigers* is true, or that $1 = 0$ is true, then the formal language or system at hand should be diagnosed and repaired.

Toward the end of the chapter, you will see that sometimes the repairs provided by changing a system are not enough to overcome some greater difficulties. The section on Kurt Gödel's Incompleteness Theorems discusses these issues. The chapter concludes with the curious Unexpected Hanging Paradox. This one is easy to grasp—an apparently harmless iceberg whose challenges exist deep beneath the surface. This paradox doesn't seem to involve self-reference at first. The logical resolution uses some of the Gödelian concepts introduced before it. An altogether different approach is then offered, focusing on beliefs and knowledge and giving you a taste of the broader class of epistemic paradoxes.

This chapter makes extensive use of set builder notation and logical notation, both of which can be found in the "Notation Guide" at the end of the book.

7.1   Russell's Paradox

Russell's Paradox was discovered independently by Ernst Zermelo in 1899 and Bertrand Russell in 1901. It is easy to conceive of a set that does not contain itself as a member. For example, the set of all pillows. If an object is a pillow, it belongs in the set. But the set itself is not a pillow, and therefore not a member of itself. Call sets like this "self-excluded."

Naively, you might also conceive of a set that does contain itself as a member. Call sets like this "self-included."

Now for the paradox. Let R be the set of all self-excluded sets—in other words, *the set of all sets that do not contain themselves*. Is R a member of itself? If not, it is a self-excluded set, and therefore belongs in R. If so, it is not a self-excluded set, and therefore does *not* belong in R. R is a member of itself if and only if it is not. This is a contradiction.

The language used to describe this paradox is unwieldy. If it hasn't made sense yet, there's a real-world example that is often used to demonstrate it. Suppose that a local librarian is tasked with creating a catalog that lists every book in her library. After she finishes listing every book, she has created a new book, that being the catalog itself. She must decide whether to list the catalog in itself.

Suppose that every librarian across the country is given this same task. Some of these librarians decide to list their catalogs in themselves, while others do not. All these catalogs are then mailed to the national library and divided into two large stacks: those that list themselves and those that do not. The national librarian is tasked with creating two master catalogs, one for each stack, each that is complete and correct (see figure 7.1).

The national librarian must decide whether to list each master catalog in itself. She begins with the master "catalog of self-included catalogs" and has no trouble. If she does self-include, then the master becomes a catalog that lists itself, and therfore its own inclusion is correct and its contents complete. If she self-excludes, then the master becomes a catalog that does not list itself, and therefore its own exclusion is correct and its contents complete. Correctness and completeness are in some sense self-fulfilling prophecies. She cannot go wrong with this master.

Now the national librarian must decide whether the master catalog of self-excluded catalogs should list itself. If she decides yes, the self-listing is incorrect because then this master is actually self-included. If she decides no,

Figure 7.1
Master cataloging at the national library.

the master is incomplete because the master is self-excluded but unlisted. Either way, she will fail in her project.

Another version of the paradox was suggested by a friend of Russell's and has become known as the Barber Paradox. Consider a barber who shaves all those, and only those, who do not shave themselves. Does the barber shave himself? He does if, and only if, he does not. This can be turned into a puzzle: "A barber shaves all the men, and only the men, who do not shave themselves. What is the gender of the barber?" The barber must be a *woman*; otherwise the situation would imply a contradiction.

Other Russell-like paradoxes can be created from a generalized version of this form, for some appropriate transitive verb: "A *verber* that *verbs* all those, and only those, that do not *verb* themselves." Among these is the Grelling-Nelson Paradox, which is as follows. The word *heterological* is defined as "inapplicable to itself." Some examples of heterological words are

- *Pulchritudinous*, an ugly word meaning "beautiful"
- *Hyphenated*, which is un-hyphenated; and *un-hyphenated*, which is hyphenated
- *Unwritten*, written here
- *Diminutive*, a big word meaning "small"
- *Unpronounceable*, which is pronounceable

Now, for the question: Is *heterological* heterological, itself? It refers to "a word that describes all words, and only words, that do not describe themselves." It describes itself if and only if it doesn't.

We now return to R, the paradoxical "set of all sets that do not contain themselves."

Claim 1
The set R either does or does not contain itself, but we will never be able to know. The only way to find out is to physically build the set, which would require infinite time and resources.

Claim 2
The set R alternates between containing itself and not, each time a Planck time interval passes. That is the unit of time required for light to travel a Planck length, which is the scale at which classical notions of spacetime cease to be valid.

Claim 3
A system of mathematics that would allow this kind of contradiction to arise must be poorly formed. If a formal set theory produces this contradiction, there must be something wrong with its axioms or the logical language used to express them.

Discussion and Resolution

Claim 3 is the correct view. Russell's Paradox emerges in naïve set theory—any formulation of set theory that is informal and capable of being expressed in natural language rather than with formal logic. In naïve set theory, the problematic set is defined symbolically as follows:

$$R = \{x \mid x \notin x\}$$

This gives us $R \in R$ if and only if $R \notin R$.

Russell came upon this paradox when considering the application of a Cantor-like diagonal argument to a "supposed class of all imaginable objects." He reflects upon this supposed class in *Introduction to Mathematical Philosophy*:

> The comprehensive class we are considering, which is to embrace everything, must embrace itself as one of its members. In other words, if there is such a thing as "everything," then, "everything" is something, and is a member of the class "everything." But normally a class is not a member of itself. Mankind, for example, is not a man. Form now the assemblage of all classes which are not members of themselves. This is a class: is it a member of itself or not? If it is, it is one of those classes that are not members of themselves, i.e., it is not a member of itself. If it is not, it is not one of those classes that are not members of themselves, i.e. it is a member of itself. Thus of the two hypotheses—that it is, and that it is not, a member of itself—each implies its contradictory. This is a contradiction.

Type theory and axiomatic set theory are two standard frameworks used to avoid Russell's Paradox.

Russell's Theory of Types

Russell realized from the discovery of this paradox certain ways in which natural language is incoherent. His ideas for addressing this problem coalesced in the ramified theory of types he proposed in 1908, which would feature prominently in *Principia Mathematica*. His ramified theory of types set out the rules that a formal ideal language must follow to avoid problems like those caused by self-reference. In this ideal language, mathematical entities and propositions, among other things, are cast into infinite hierarchies of types. Objects of any given type can be constructed only from objects of lower-level types, but never from the same level or higher levels.

We can apply that theory to the formation of sets. We start with some specified collection of elements. From these, we can construct sets of elements that exist at a level higher than the mere elements. Then we can create sets of sets of elements. A set at level $n + 1$ can be created from objects at level n. But no set at level $n + 1$ can contain objects at level $n + 1$ or greater. A crude analogy: One might construct a castle from pebbles, but it is meaningless to speak of constructing a pebble from castles. This solution is designed to avoid vicious circles, as described by Russell and Alfred Whitehead in *Principia Mathematica*:

> An analysis of the paradoxes to be avoided shows that they all result from a kind of vicious circle. The vicious circles in question arise from supposing that a collection of objects may contain members which can only be defined by means of the collection as a whole. Thus, for example, the collection of *propositions* will be supposed to contain a proposition stating that "all propositions are either true or false." It would seem, however, that such a statement could not be legitimate unless "all propositions" referred to some already definite collection, which it cannot do if new propositions are created by statements about "all propositions." We shall, therefore, have to say that statements about "all propositions" are meaningless. ... The principle which enables us to avoid illegitimate totalities may be stated as follows: "Whatever involves *all* of a collection must not be one of the collection"; or, conversely: "If, provided a certain collection had a total, it would have members only definable in terms of that total, then the said collection has no total." We shall call this the "vicious-circle principle," because it enables us to avoid the vicious circles involved in the assumption of illegitimate totalities.

If "all sets that do not contain themeslves" were put into a collection R, that collection would exist at a higher level than its elements and could not be eligible for membership as an element in itself. In the language of Russell's logic, it would be ungrammatical to speak of R as being a member of itself.

Axiomatic Set Theory

The more commonly accepted solution to this paradox is to lay out a rigorous set of axioms that form the foundation of a more consistent set theory. One such axiomatic system is Zermelo-Fraenkel (ZF) set theory, which was designed specifically to avoid Russell's Paradox and others like it. (There is a historically controversial axiom called the *axiom of choice*. ZF set theory, together with the axiom of choice, is called *ZFC set theory*, with C meaning "choice." ZFC is now the standard axiomatic set theory used in mathematics.) There are other axiomatic set theories, but here we focus on the resolution provided by ZF.

In naïve set theory, sets are constructed as "all the things that satisfy some predicate." In other words, "all the x such that $\phi(x)$ is true," where $\phi(x)$ is a predicate that can be true or false for different values of x. Using set builder notation, naïve set theory lets us build a set S like this:

$$S = \{x|\phi(x)\}$$

and so $R = \{x|x \notin x\}$ is permitted because we can define $\phi(x)$ as the predicate "$x \notin x$."

This would not be legal in ZF, however. ZF includes an axiom called the *axiom of specification*, which prohibits us from saying "all things that." ZF requires that we build sets in a more concrete fashion, starting with choosing candidate elements from some collection that we know is a valid set. Instead of saying "all things that," we must say, "all members of set A that," where A is a known legal set. In set builder notation, this means that the only way to construct a valid set S is

$$S = \{x \in A|\phi(x)\}$$

We start with some valid set A, then create S as a subset of A by taking all of A's elements that satisfy the predicate $\phi(x)$. Valid sets like A can be built using basic building blocks that ZF allows, like null sets, and performing operations that ZF allows to make new valid sets, like unions and power sets.

Returning to Russell's Paradox, suppose we followed ZF's rules and collected "all members of A that are not members of themselves." In set builder notation:

$$R = \{x \in A|x \notin x\}$$

This is perfectly legal, and there is no contradiction.

Now let's return to the cataloguing application from earlier in this chapter. Zermelo and Fraenkel would never give the librarian such an open-ended task as, "Create a master catalog of all catalogs that exclude themselves." That's how problems arise. What they *would* say is, "Create a master catalog from only among the catalogs stacked here, on this table, of those which exclude themselves."

Set, Class, and Proper Class

The progression from naïve set theory to axiomatic set theory gave rise to the concepts of classes and proper classes, both distinct from sets:

- A *class* is any collection of things that have a common property.
- A *set* is a class contained in some class.
- A *proper class* is a class that is not a set.

Let R be the collection of "all sets that do not contain themselves." R is a class because it is a collection of things (sets) that have a common property (do not contain themselves).

Is it a set? Axiomatic set theory tells us that R must contain only elements that are members of some class A, where

$$R = \{x \in A \mid x \notin x\}$$

If R were a member of A, then a contradiction would arise: R would be a member of itself if and only if it were not. Therefore, R cannot be a member of any class A. That means R is not a set.

Because R is a class but not a set, it is a proper class.

7.2   The Liar Paradox Family

The Liar Paradox is among the most famous of paradoxes. Philosophers, logicians, and thinkers in a great many fields have pondered over this one—and its variations, which form a rich and interesting family of Liar Paradoxes—for more than two millennia. The Liar Paradox began historically with the Greek philosopher Epimenides of Knossos, who lived circa 600 B.C. Epimenides, himself a Cretan, made this famous claim:

All Cretans are liars.

For historical accuracy, it should be noted that Epimenides probably did not intend his statement in a paradoxical sense. The phrase appeared in one of his poems, in which he called Cretans "always liars" for denying the immortality of Zeus. By "liar," he probably meant someone who speaks falsehoods often, but not necessarily exclusively. But under a strict definition of the word, there's a puzzling irony to Epimenides' claim. Let a "liar" be someone who always speaks falsely, and a "truth-teller" someone who always speaks truthfully.

Suppose the statement is true. The statement was spoken by a Cretan who, according to the statement itself, must be a liar. But no liar can utter the truth, so anything Epimenides says must be false. And if it is false that "All Cretans are liars," then all Cretans must be truth-tellers, which contradicts the original statement (and therefore itself). It seems that a contradiction inevitably arises from Epimenides' claim.

But it only seems that way. There is a flaw in the logic given here, and thus a solution to this particular puzzle. It is correct that the statement cannot be true; its truth would imply a contradiction. But if it is false that "All Cretans are liars," then it is not necessarily true that all Cretans are truth-tellers. The negation of "All Cretans are liars" is "Not all Cretans are liars," or equivalently, "There is at least one Cretan who is not a liar." Furthermore, someone who is not a liar does not have to be a truth-teller. A person who sometimes speaks falsely and sometimes speaks truthfully is neither a liar nor a truth-teller as defined.

There is no contradiction if at least one Cretan exists who is a nonliar. We would simply say to Epimenides, "Your statement is false." No problem.

Unless. What if there were only one Cretan: Epimenides?

Classical Liar

If Epimenides were the only Cretan, we wouldn't have such an easy way out. It would be as if Epimenides were saying, "Everything I say is false," or "I am lying to you right now," which are both forms of the so-called Classical Liar Statement:

This statement is false.

If the statement is false, then it must be true. But if it's true, it must be false. If we try to give this statement a binary truth value, we face a contradiction. The liar phenomenon occurs in many variants, some of which are described next.

Cycles

Consider the following pair:

The next statement is false.

The previous statement is true.

If the second statement is false, then the first statement must be false. If the first statement is false, the second statement must be true. This would imply the first statement is true, and the second statement is false. The truth of either statement implies its own falsity, and vice versa—a contradiction, even though explicit self-reference was avoided (though not circular reference). Longer cycles of finite length can be constructed.

This construction is equivalent to the Card Paradox, conceived by the British logician Philip Jourdain. Jourdain wrote a sentence on each face of a card. One claimed the other to be false, and one claimed the other to be true. Another variant is the so-called No-No Paradox, constructed by a pair of sentences, each claiming the other to be false.

Strengthened Liar

The *negation* (or opposite) of truth is falsity. The *denial* of truth is untruth; to deny that something is true is to say it is not true. Are falsity and untruth the same concept? Not quite, if you reject the principle of *bivalence*, which holds that every sentence is strictly true or false.

Consider the possibility of a third alternative to truth and falsity. If a statement can be true, false, or K, then the statement is not true if it is either false or K. Here, K can be understood to mean nonsensical, meaningless, incoherent, or some other alternative to the true/false binary. Falsity implies untruth, but untruth does not imply falsity, since K is also an option. With this in mind, consider the so-called Strengthened Liar Statement:

This sentence is not true.

Surely the sentence is either true or not true. If it is true, then its meaning says it is not true—a contradiction. If it is not true, its meaning is shown to be true—another contradiction. We are still stuck in the paradox. We can also formulate the Strengthened Liar's Statement like this:

This sentence is false or K.

If the statement is true, then its meaning implies it is either false or K—a contradiction. If the statement is false, then it is neither false nor K—another contradiction. If it is K, then the statement itself is true—still another contradiction.

Some philosophers attempt to resolve paradox by calling the Classical Liar's Statement meaningless, but no such resolution can be sufficient to handle the Strengthened Liar. "Meaningless" is one of any number of interpretations K can have.

Contingent Liar

It is possible for the existence of a paradox to rely on contingent facts. Consider the statement:

It is raining, and this sentence is false.

If it is not actually raining, then the sentence is nonparadoxically false. But if it is actually raining, the paradox is present. Here, the factual state of the weather, an empirical fact, determines whether the statement is paradoxical or not.

Yablo's Paradox

The philosopher Stephen Yablo devised a liar-like paradox, which he argues avoids self-reference and circularity, although it is debated whether this is true. Yablo's Paradox works like this. Imagine an infinite series of ordered statements, each declaring every subsequent statement to be false:

P_1 Every subsequent statement is false.
P_2 Every subsequent statement is false.
P_3 Every subsequent statement is false.
$\vdots$ $\vdots$

Is P_1 true or false? If it is true, then every "greater" statement is false, including P_2. But if P_2 is false, then there exists some statement greater than P_2 that is true, contradicting P_1. Hence, P_1 cannot be true. Conversely, if P_1 is false, then there is some greater statement P_k that is true. But if that were the case, we could imagine creating a new list starting at P_k, and as we have already shown, the first statement on the list cannot be true.

How to Prove Anything

Anything can be proven from a contradiction. Here is one way to use a liar-like statement to prove anything. Consider the following proposition P:

Either this proposition (P) is not true, or X.

Here, X is some proposition. Suppose X is not true. By P, the only alternative is "this proposition (P) is not true," which means P is true—the classic contradiction. It must mean, therefore, that X is true; this is the only way the previous contradiction can be resolved. This reasoning holds for any statement X. It could be that X is "One equals zero," or "Aliens built the Pyramid of Giza." In fact, the fourteenth-century French philosopher Jean Buridan used this method in an attempt to prove the existence of God.

Truth-Teller

Consider the so-called Truth-Teller's Statement:

This sentence is true.

Is this statement true or false? Suppose it is true. Indeed, the statement affirms itself as true. Suppose it is false. Indeed, the statement speaks falsely, claiming to be true. The Truth-Teller's Statement is consistent and self-affirming whether we assume it to be true or false.

In the earlier section on Russell's Paradox, an application was presented describing a national librarian tasked with creating two master catalogs: all self-included catalogs, and all self-excluded catalogs. Whether or not the librarian listed each catalog in itself, the librarian was certain to succeed at creating the first catalog, and doomed to fail at creating the second. Analogously, assigning truth or falsity to the Truth-Teller's statement produces inevitable consistency, whereas doing the same for liar statements produces inevitable contradiction.

Quine's Paradox

A version of the Liar's Paradox attributed to Willard Van Orman Quine is the statement:

> "Yields falsehood when preceded by its quotation" yields falsehood when preceded by its quotation.

This sentence is awkward and difficult to parse. Try reading it a few times for it to make sense to you. The second half of the sentence is preceded by a quotation of itself. The sentence itself says that such an occurrence yields falsehood. But if it is false that it yields falsehood, then it yields truth, which means it yields falsehood, and so on.

Quine's sentence makes no explicit reference to itself as a whole. However, the possessive pronoun "its" in the second half of the sentence refers to the string of letters that gets repeated twice, one of which contains that very instance of "its."

Understanding the Problem

What's really going on with the Liar Paradox and all its variants? How do we define the problem? It may be tempting to dismiss liar statements as mere contradictions. A contradiction is a statement that asserts something to be so and not so, like "Mount Everest has snow, and Mount Everest does not have snow." We would expect language to be capable of expressing falsities such as contradictions.

We would not, however, expect language to generate or support the emergence of contradictions from its own grammatically valid sentences. That's precisely what's happening with the family of Liar Paradoxes, and the reason why we need a systematic approach to answering some broader questions, like: What properties should a coherent language satisfy? How should truth be conceived and defined? What criteria must be satisfied for statements to have a truth value? How can these answers be integrated into broader theories of syntax and truth?

Claim 1

The Classical Liar Statement is true.

Claim 2

The Classical Liar Statement is false.

Claim 3

The Classical Liar Statement is neither true nor false.

Claim 4

The Classical Liar Statement is both true and false.

Claim 5

Depending on how it is interpreted, the statement can be either true or false (one or the other, but not both) without contradiction.

Discussion and Resolution

This problem is profound, and even after thousands of years of analysis, there is no consensus on its resolution. The purpose of this discussion is to explore and critique some (but not all) of the better-known approaches to resolving the paradox and to develop a deeper understanding of truth and falsehood along the way. The discussion, alas, does not begin with an assertion that any of the claims above is the right one. That's not to say there isn't a best or correct treatment to the problem. In what follows, various arguments will be presented that support some of these claims.

Implied Assertions of Truth

Jean Buridan informally suggested an idea that has been championed and/or further developed by Charles Sanders Peirce, Arthur Prior, Eugene Mills, Neil Lefebvre, Melissa Schehlein, and others. His idea was that every statement implicitly asserts its own truth. If someone says, "Mount Everest has snow," he or she does not mean, "It may be untrue that Mount Everest has snow." He or she is indeed asserting that "It is true that Mount Everest has snow," or alternately phrased, "This statement is true, and Mount Everest has snow."

In the case of the Classical Liar, the statement is really saying "This statement is true, and this statement is false." That is a contradiction, and all contradictions are false.

But wait. You might say that the falsity of the statement vindicates the statement itself. We proved by contradiction that the statement is false, which it claims to be, so it must be true—and here we are, back in the same deadly cycle.

We can resolve this issue by realizing that the Classical Liar's Statement is not just asserting its own falsity; it is asserting both its own truth and falsity. But a contradiction can only be false. The statement is not true *and* false, as it claims to be. It is *only* false. This reasoning is consistent with Claim 2.

Verifiability

Since we are soon to be involved in assigning truth values, we should first identify criteria that a statement must meet to qualify for a truth value. Is every statement either true or false? Or are there some statements that are neither, that have a so-called *truth-value gap*?

Let's start by addressing *plain* statements, which we will define informally as those that make no mention of the concept of truth. Under what conditions can we evaluate plain statements as true or false? A plain statement has a truth value if its empirical factuality is verifiable by means of observation, measurement, and/or logical integration. For example, "Mount Everest has snow" is a statement we can test with our senses. We know the precise earthly geographical feature that is meant by "Mount Everest," we know what "snow" is, and we can observe whether there is snow on Mount Everest. The statement is verifiable.

A plain statement may fail to be verifiable, and hence have a truth-value gap, if it meets the following criteria:

- It makes a *categorical mistake*. That is, it applies a property to something to which that property does not apply. For example, "Happiness is tall." Physical height is not related to emotion.

- It uses imprecise or subjective language. For example, "The lake is deep." What counts as deep? Three meters? Fifty meters? Or, "Monet's paintings are beautiful." How is beauty defined?

- It uses words or phrases with empty or nonexistent referents. For example, if a bald man were to say "The hair on my head is brown," he would be making reference to nonexistent hair; his statement would be neither true nor false. What hair? (If a bald man were to say, "I have brown hair on my head," this would indeed be a false statement, but with no empty reference.) Statements with references to fictional entities may contain truth-value gaps if they treat those entities as if they were real.

- It makes a declaration about any aspect of the future that cannot be foretold with certainty. For instance, "Tomorrow it will rain." The accuracy of this prediction will become known in a day's time, but at the moment the statement is made, its truth value cannot be known.

The literature on such conditions and others, and on truth-value gaps, is vast and without consensus. But we can draw a basic idea without getting lost in details: For a plain statement to have a truth value, it must have ascertainable consistency or inconsistency with reality.

It is possible for self-referential statements to be verifiable. Consider the following sentence:

This sentence is in English.

It's not immediately clear what is meant by "This sentence." We want to replace that ambiguous phrase with a concrete referent. We can do so by generating a new sentence from the original:

"This sentence is in English" is in English.

The ambiguity disappears when we replace the label with something we can interpret. We can now evaluate the sentence within the quotation marks as being in English and assign a value of *true*.

Let's consider the Classical Liar's Statement, however:

This statement is false.

Is this statement verifiable? Let's try to clear up all ambiguities, as before. What exactly is meant by "this statement"? The label has no interpretation yet. In an effort to replace every ambiguous term with a concrete referent, we can generate a new sentence by replacing "This statement" with the entire original:

"This statement is false" is false.

But now since we must determine the truth value of the sentence in quotations, we must make sure all its references are concrete, and so we must disambiguate further by generating:

"'This statement is false' is false" is false.

Then we must generate another statement, and another. Each time a new sentence is generated, that new sentence will contain an ambiguous term. For any arbitrarily large n, the nth sentence will be unverifiable. No matter how far we unpack the original sentence, the problem of verifiability will persist.

So, why don't we just rephrase the original sentence in a way that eliminates the problematic ambiguity? For example, we could say instead:

I am lying right now.

Does that resolve the ambiguity? It may seem like it, but "right now" is arguably ambiguous. What point in time is it referencing? Most likely, it is the moment when the sentence was uttered. So we can generate a new sentence:

I am lying when I say, "I am lying right now."

"Right now" appears again, so we generate the next level:

I am lying when I say, "I am lying when I say, 'I am lying right now.'"

Once again, the problem continues into infinite regress. Can we eliminate the problematic time stamp of "right now"? What if we just say:

I am lying.

Now it's less clear that the speaker is referring to that very sentence. Is the speaker lying about something else, or about that statement? The idea behind this argument is that liar-like statements in any form are unverifiable and therefore have no truth value.

Russell's Theory of Types

Recall Bertrand Russell's *theory of types*, developed in response to Russell's Paradox. (See the section entitled "Russell's Paradox," earlier in this chapter; the *theory of types* is described in the discussion and resolution.) Russell addresses the Liar Paradox directly in *Mathematical Logic as Based on the Theory of Types*. His logic works like this. Consider the statement:

> I am lying.

This statement can be interpreted as.

> There is a proposition that I am affirming and that is false.

Notice this previous statement begins with "There is." Statements of this form deny that the opposite is always true; for instance, "There is a mountain that is tall" is equivalent to "It is not true that every mountain is not tall." We can perform this same inversion on the previous sentence to generate the equivalent. The negation of the condition "which I am affirming and which is false" is "which I am either not affirming or which is true." The inverted but equivalent sentence becomes

> It is not true of all propositions that either
>
> I am not affirming them or they are true.

This is equivalent to

> It is not true for all propositions p that if I affirm p, p is true.

This is itself a proposition which the speaker is affirming, and therefore falls under the scope of "all propositions p that I affirm." Russell argues that the use of "all propositions" is illegitimate. If it were legitimate to make statements about "all propositions," then one could assert a new proposition—which would be one of "all propositions"—that did not accurately describe itself. Russell writes, "Whatever we suppose to be the totality of propositions, statements about this totality generate new propositions which, on pain of contradiction, must lie outside the totality. It is useless to enlarge the totality, for that equally enlarges the scope of statements about the totality. Hence there must be no totality of propositions, and 'all propositions' must be a meaningless phrase."

The problem is solved, Russell argues, by the *theory of types*:

> Whenever "all propositions" are mentioned, we must substitute "all propositions of order n," where it is indifferent what value we give to n, but it is essential that n should have some value. Thus when a man says "I am lying,"

we must interpret him as meaning: "There is a proposition of order n, which I affirm, and which is false." This is a proposition of order $n + 1$; hence the man is not affirming any propositions of order n; hence his statement is false, and yet its falsehood does not imply, as that of "I am lying" appeared to do, that he is making a true statement. This solves the liar.

Russell makes the case that "I am lying" is a meaningless statement because it relies on a concept of "all propositions" that is meaningless. By casting a liar statement into a hierarchy, whereby the liar makes a proposition at level $n + 1$ about some level n—but claims to do so at level n—we see that the liar is indeed lying, but there is no contradiction.

Russell believed that many logical paradoxes could be attributed to violations of this principle: Whatever describes a type cannot be a member of that type. His resolution also handles Yablo's Paradox, whose statements can be cast into an infinitely descending hierarchy, by requiring that the hierarchy have some lowest level (which Yablo's statements do not) that serves as the foundation for those higher up.

His solution, however, places serious restrictions on what can be considered a syntactically well formed sentence. For example, the sentence mentioned earlier in the chapter, "This sentence is in English," is one we showed to be perfectly verifiable, but it has no place in Russell's hierarchy; it would be considered illegitimate and ungrammatical. His system also makes it impossible to describe, with well-formed sentences, certain aspects of the system itself.

Tarski's Theory of Truth

Alfred Tarski set out to address the paradox using hierarchies, but in a less restrictive manner. He wanted to develop a formal language that would preclude the possibility of liar-like contradictions, but not self-reference altogether. His idea was to make his language as close to natural language as possible.

Some terminology will be helpful. Remember that a predicate is a part of a sentence that contains a verb and says something about the subject. In the sentence "Mount Everest has snow," the subject is "Mount Everest" and the predicate is "has snow." In logic, the *truth predicate* is the predicate "is true" as applicable to other statements. For example, "'Mount Everest has snow' is true" uses the truth predicate to assign a value of truth to the statement in single quotes. A language is said to have its own truth predicate if it permits statements that use the truth predicate to assign truth value to statements in the same language. In other words, such a language can express whether sentences in the same language are true or false. A language that allows both "Mount Everest has snow" and "'Mount Everest has snow' is true" to be expressed has its own truth predicate. Tarski calls such languages *semantically closed*.

Tarski argued that the Liar Paradox can arise only in semantically closed languages. His way out of the paradox was to envision a hierarchy of languages, such that any given language can predicate truth or falsity of statements only in lower-level languages. A language could never predicate truth of sentences in the same or a higher level. The lowest language in the hierarchy, which he calls the *object language*, permits no truth predicates at all. A higher language that does permit truth predicates is called a *metalanguage*. The statement "Mount Everest has snow" would exist in the object language, and "The sentence 'Mount Everest has snow' in the object language is true" would be expressed in the metalanguage.

"This sentence is false" is illegitimate under Tarski's system. The sentence makes reference to a sentence (itself) in the same language level and is therefore ungrammatical. Tarski's language hierarchy also precludes the possibility of paradoxical chains, such as "The next sentence is true" and "The previous sentence is false." Each must exist at a higher level than the other, which is impossible, so the possibility of this kind of chain is ruled out.

In 1936, Tarski proved a result called the Undefinability Theorem. Informally, the theorem states that no formal language can define its own concept of truth without producing semantic incoherence. There is no consistent formal language that can define its own truth predicate. A key step in the proof relies on analysis of the liar's sentence. It is possible, however, for a concept of truth in an object language to be defined by a metalanguage. The hierarchy of languages can be seen as a practical solution to the problems and impossibilities proved by Tarski's Undefinability Theorem.

There are some drawbacks to Tarski's hierarchy. Philosopher and logician Saul Kripke points out a few. For one, certain nonparadoxical statements are precluded from being expressed at all. Suppose each of two brothers wants to say of the other, "Everything my brother says is true." There would be nothing paradoxical about the brothers saying so. It would be desirable for a system of language to permit these simultaneous statements. But under Tarski's rules, each such statement would have to exist at a higher level than the other, which is impossible, so in Tarski's system, no such pair of utterances would be legitimate.

Kripke also points out the failure of Tarski's solution to handle "contingent" versions of the Liar Paradox, like the one seen earlier: "It is raining, and this statement is false." This sentence is paradoxical if and only if it is raining; otherwise, it is nonparadoxically false. Sentences like that cannot be expressed in Tarski's system, even when the contingent facts determine them to be nonparadoxical.

Tarski's system requires statements to have an intrinsic level fixed in advance, but a speaker may find it difficult or impossible to know what level to assign a given statement. For example, if a speaker wished to remark about most or all of what his or her friend had uttered, who had uttered sentences made by another friend, and so on, our speaker would have a tough time picking the appropriate level for his or her own language.

Tarski's system also precludes the ability to address truth value at every level of the hierarchy. For example, the statement "Every level of language has some true sentences" is ungrammatical because it must be higher than every level of the hierarchy, which is an impossibility unless transfinite levels are allowed. It would seem desirable to be able to express global truth predicates about the hierarchy as a whole, but such a language would exist outside the finite hierarchy.

This is quite a heavy restriction. If transfinite levels are forbidden, how could one teach the nature of Tarski's system while following its rules? His language would not be able to say everything that is true about itself. And if transfinite levels are allowed, Kripke argues, a host of other nontrivial problems are introduced.

Kripke's Theory of Truth

Recall our informal definition earlier in this chapter of a *plain* sentence as one that makes no mention of the concept of truth. Of course, not all statements are plain. We can make statements that mention the concept of truth, in particular by asserting other statements to be true or false. For example:

> The statement "Mount Everest has snow" is true.

One could then speak of that sentence as true, and so on, forming a chain. In his theory of truth, Kripke introduces a concept of grounding, presented here in a nontechnical fashion: If a chain terminates in a sentence that is plain and verifiable, then it is *grounded*. Otherwise, it is ungrounded, and ungrounded statements, says Kripke, fall into the truth-value gap.

One possible resolution to the paradox, which is consistent with Claim 3, is that liar statements are ungrounded and therefore have no truth value. Groundedness, Kripke would argue, is a conceptual precursor on which valid concepts of truth and falsehood depend. Recall ZFC's resolution to Russell's Paradox: ZFC requires that sets must be constructed as subsets from existing sets that are known to be valid. That statements should be grounded is an analogous criterion.

This is one of the fundamental ideas in the formal structure of Kripke's broader theory of truth, published in 1975. His theory differs from Tarski's in that it permits a language to have its own global truth predicate. (Robert Martin and Peter Woodruff published a theory in 1975 that allows the same.) Kripke avoids the limitations imposed by Tarski's Undefinability Theorem, which applies only to languages whose sentences are strictly true or false. The classification of some sentences as ungrounded enables Kripke to prove the existence of a semantically coherent language that defines its own concept of truth.

Barwise and Etchemendy's Treatment

Jon Barwise and John Etchemendy present a resolution that entails keeping track of all the relevant facts in stages through the argument. A collection of facts is called a *situation*, and the situation changes as the reasoning proceeds. They conclude that the Liar's Statement is ambiguous. It is either true or false, consistent with Claim 1 or 2, depending on whether the Liar's Statement is interpreted as denying or negating itself. Understanding Barwise and Etchemendy's treatment entails learning a new framework and notation, which are omitted here but explained in their work *The Liar: An Essay on Truth and Circularity*.

Self-Contradictory Systems

A multitude of contexts can be conceived that resemble the Liar Paradox and that follow a similar structure of producing a self-contradictory system. Consider the following scenarios:

- Among the best known is the paradox of Buridan's Bridge, conceived by Jean Buridan. As the story goes, Plato guards a bridge over a river that Socrates wishes to cross. Plato says to Socrates,

"If the first statement you utter in reply to me is true, I will permit you to cross. If it is false, I will throw you into the water." Socrates replies, "You will throw me into the water." Plato's brain then freezes. He doesn't know what to do. If he throws Socrates into the water, he will have proven Socrates' statement to be true, hence violating his own promise to let Socrates cross the bridge in such a case. But if he lets Socrates cross the bridge, he will have proven Socrates' statement to be false, hence violating his own promise to throw Socrates into the water in such a case. (We are to assume that Plato cannot throw Socrates into the water once he has crossed the bridge.)

- The Pinocchio Paradox was proposed in 2001 by then eleven-year-old Veronique Eldridge-Smith. Pinocchio is a puppet whose nose grows whenever he utters a lie. Suppose Pinocchio were to say, "My nose will now grow." If this were true, his nose would not grow; but then it would be a lie, causing his nose to grow; but then it would be true. What would happen if Pinocchio made such an utterance?

- There is a famous problem in computability theory called the *halting problem*. It is stated as follows: Can an algorithm be written that looks at the code of any program and outputs whether that program will run forever or eventually terminate (halt)? Alan Turing proved that there is no such algorithm. He did so by proposing and analyzing a program that essentially works like this: "If this program halts, let it loop forever; otherwise, say 'it doesn't halt' and terminate." Contradiction ensues.

- A general says to his troops: "Obey my next order. Disobey my last order." Can the troops follow these orders? (It also can be expressed as one sentence: "I command you to disobey this order.")

- You say to a genie who promises to grant two wishes, "I wish that you will grant my next wish. I wish that you will not grant my previous wish." Can the genie grant your wishes? (Expressed in one sentence: "I wish that this wish will not be granted.")

- Suppose you were to play a game with someone. The winner is the least competitive player, where competitiveness is determined by who's winning. If you try to win, you're being competitive, which puts you in a losing position. Losing makes you less competitive, and therefore puts you in a winning position. Would you try hard to win?

See if you can dream up any scenarios that resemble those given here. What do they have in common? Much has been written about several of them, particularly the first three. Attempted resolutions often deal with the subtleties of each particular problem—for example, whether claims about the future can have a truth value. The idea is not to delve into the subtleties of each case here, but rather to point out a structural similarity among them. We can model each of these systems with two *state variables*, each of whose value depends on the other, but in mutually inconsistent ways.

Consider the following system of equations:

$$x = \begin{cases} 0, & \text{if } y = 1 \\ 1, & \text{if } y = 0 \end{cases}$$

$$y = \begin{cases} 1, & \text{if } x = 1 \\ 0, & \text{if } x = 0 \end{cases}$$

By design, there is no pair (x, y) that satisfies both these equations. It makes no sense to ask what would happen if we asked this system to settle on some stable values for (x, y). Such values do not exist. In the same way, it makes no sense to ask what happens when Socrates replies as he does, or when Pinocchio says his nose will grow, and so on. There is no physical or mathematical system that can follow the rules as described. Any such rules would be broken by necessity.

What makes this reality psychologically difficult to accept? The rules are introduced in ways that make them seem harmless. Consider the case of Buridan's Bridge. Plato's promise seems like one that he can keep no matter what, independent of what Socrates says. Then Socrates does something that is entirely physically possible: He makes an utterance. It's not until further analysis that we realize Plato's ability to keep his own promise depends on the nature of Socrates' statement. Socrates speaks in such a way as to cause an interaction between the two, defining a system of rules that cannot be followed. The scenarios described can be cast into the state-variable model like this:

- In the case of Buridan's Bridge, one state variable x is the truth or falsity of Socrates' claim. The other state variable y is the action taken by Plato to throw him into the river or let him cross.

- In the Pinocchio Paradox, one state variable x is the truth or falsity of Pinocchio's claim. The other state variable y is the growth or nongrowth of his nose.
- In the halting problem, one state variable x is the output of the subroutine that asks whether the program will halt or not. The other state variable y takes a value of yes or no, determined by whether the program loops forever or halts.
- In the problem of the general's orders, one state variable x reflects whether the first order is obeyed. The other state variable y reflects whether the second order is obeyed.
- In the genie problem, one state variable x reflects whether the first wish is granted. The other state variable y reflects whether the second wish is granted.
- In the game of least competitiveness, one state variable x reflects whether you are winning. The other state variable y reflects your level of competitiveness.

7.3 Berry's Paradox

Bertrand Russell described this paradox in 1908:

> The number of syllables in the English names of finite integers tends to increase as the integers grow larger, and must gradually increase indefinitely, since only a finite number of names can be made with a given finite number of syllables. Hence the names of some integers must consist of at least nineteen syllables, and among these there must be a least. Hence "the least integer not nameable in fewer than nineteen syllables" must denote a definite integer; in fact, it denotes 111,777. But "the least integer not nameable in fewer than nineteen syllables" is itself a name consisting of eighteen syllables; hence the least integer not nameable in fewer than nineteen syllables can be named in eighteen syllables, which is a contradiction.

Russell attributes the paradox to G. G. Berry, a librarian at the Bodleian Library at Oxford University.

Claim 1
What counts as a "syllable" must be ambiguous.

Claim 2
The contradiction arises from an illegitimate concept sneaking in—the concept of "all names."

Discussion and Resolution
Recall from the section on Russell's Paradox earlier in this chapter the rules Russell specified in relation to the vicious circle principle:

> "Whatever involves all of a collection must not be one of the collection"; or, conversely: "If, provided a certain collection had a total, it would have members only definable in terms of that total, then the said collection has no total."

This rule, Russell writes, has been broken here:

> "The least integer not nameable in fewer than nineteen syllables" involves the totality of names, for it is "the least integer such that all names either do not apply to it or have more than nineteen syllables." Here we assume, in obtaining the contradiction, that a phrase containing "all names" is itself a name, though it appears from the contradiction that it can not be one of the names which were supposed to be all the names there are. Hence "all names" is an illegitimate notion.

In the spirit of Russell's theory of types, the paradox can be resolved by stratifying "nameability" into a hierarchy of types. The expression "the least integer not level-n nameable in fewer than nineteen syllables" is an expression of nameability level $n+1$. The concept of "all names" has no meaning unless tied to a certain level, in which case "all names" exists a level higher. In this framework there is no contradiction.

7.4   Richard's Paradox

Richard's Paradox was conceived by the French mathematician Jules Richard in 1905. Consider the set that contains each decimal number that can be unambiguously defined in finite words. Call this set X. Here are some sample unambiguous phrases that define numbers this set will therefore contain:

- Zero point five seven
- The square root of zero point eight

- Zero, followed by a decimal point, followed by an infinite and alternating repetition of the numbers two and three, beginning with two.

The result of this set is an infinite list of real numbers.

Let us define another number, r, as follows. Arrange all numbers in X by increasing length of their definitions; then alphabetize each set of definitions of equal length. Now construct r such that its nth decimal is 0 if the n decimal place of the nth number on the list is not 0; otherwise, let the nth decimal be 1.

Note that this is the key method used in Cantor's diagonalization argument, explained in chapter 1, used to prove no bijection exists between countable and uncountable infinities.

This method generates a number that cannot be a member of the initial list, because it differs from every number on the list in at least one decimal place. Hence, $r \notin X$. Yet the number r was defined unambiguously in finite words, and therefore is a member of X. We have a contradiction.

Claim 1

It must be that r is ambiguously defined and is therefore a member in X. The diagonalization argument must not apply here.

Claim 2

The set X is constructed illegitimately. It uses a notion of "all definitions," but the meaning of some of those definitions is derived from "all definitions."

Discussion and Resolution

Claim 2 is correct. This paradox falls within the family of paradoxes resolved by Russell's theory of types or axiomatic set theory. "All definitions" is not a legitimate concept because the supposition of its legitimacy gives rise to definitions outside the totality of "all definitions," on pain of contradiction. It would be valid to speak of a set of definitions at some level n and create new definitions of level $n+1$ based on the totality of level-n definitions.

7.5   Burali-Forti Paradox

The Burali-Forti Paradox invokes a concept you will need to understand thoroughly before proceeding. That is the concept of ordinal numbers. Readers familiar with ordinals may comfortably skip this primer section.

Primer on Ordinal Numbers

In chapter 1, in the book, the concept of cardinality was introduced. A cardinal number expresses the size of a collection. Numbers also can be used to express position in an ordered sequence, like "first," "second," "third," and so on. These numbers are called *ordinal* numbers, or just *ordinals*.

A *well-ordered* set must have a relation, $\prec$, that meets three criteria. First, for any elements x and y in the set, either $x \prec y$, $y \prec x$, or $x = y$ (exactly one of these is true). Second, the relation must be transitive. Third, the set must contain a *least element* according to that relation—an x such that there is no $y \prec x$.

We can now proceed to an intuitive understanding of ordinal numbers, followed by a formal definition. What exactly is an ordinal number?

Imagine a series of rocks lined up from left to right. Suppose we assign positional order by starting at the leftmost rock and working to the right, so that each rock is said to have a greater position than any rock to its left. This set of rocks is well-ordered. Now, how can we identify specific rocks without using words like "first," "second," "third," et cetera? The rock we would call "first" is uniquely identified as that which has no rocks to its left. The "second" rock is uniquely identified as the rock to the left of which is a rock that has no rocks to its left. The "third" rock is uniquely identified as the rock to the left of which is a rock to the left of which is a rock that has no rocks to its left. And it goes on.

Notice each such rock's position is described uniquely by referencing all rocks to its left. This way of identifying positions in a well-ordered set gives us a formal way of defining ordinal numbers. John von Neumann defines an ordinal number as *the well-ordered set of all preceding ordinal numbers*. Nothing comes before what is first (there is no zeroth place in a race), so the first ordinal is defined as the set containing nothing—the set containing the null set. Subsequent ordinals build from there:

Ordinal	Definition	Breakdown
0	{}	∅
1	{0}	{∅}
2	{0, 1}	{∅, {∅}}
3	{0, 1, 2}	{∅, {∅}, {∅, {∅}}}
⋮	⋮	⋮

It will help to understand the concept of ordinal numbers if we start counting them out in order. The ordinals begin with 1 (indicating first), 2 (second), 3 (third), and continue unendingly in this manner throughout the natural numbers. Can there be an ordinal greater than any of these? It turns out that there can. Our definition of ordinals gives us the ability to transcend infinity.

Remember that an ordinal is simply the well-ordered set of all its predecessor ordinals. In this case, we treat all the natural-numbered ordinals as predecessors and put them in a well-ordered set. That set defines a new ordinal, the least ordinal greater than any of the natural-numbered ordinals. This ordinal is the first transfinite ordinal and is denoted by ω, the Greek letter omega. Here, ω is said to be a limit ordinal. It does not have an immediate predecessor, as there is no largest natural number. In the same way that there is no "greatest real number smaller than 1," there is no "greatest ordinal smaller than ω."

But it does have a successor, so we can keep counting. The successor to ω is simply $\omega + 1$, then $\omega + 2$, and so on. After another countable infinity of successors comes the next limit ordinal $\omega \cdot 2$, then $\omega \cdot 2 + 1$, $\omega \cdot 2 + 2$, and so on. New products $\omega \cdot m$ are reached until eventually m achieves transfinite status and ω^2 is reached. Eventually, the exponent will achieve transfinite status with ω^ω, and as ever more infinities are transcended, a ladder of exponentiation:

$$\omega^{\omega^{\omega^{\cdot^{\cdot^{\cdot}}}}}$$

And so the proceeding into ever-higher ordinals continues.

It may take some time to get comfortable with the concept of ordinals. Concepts like "first," "second," and "third" are intuitive, but generalizing the concepts of position and order, and formalizing the definition of ordinality, are not so easy. It is important to understand the concept of ordinality before reading the Burali-Forti Paradox.

The Paradox

The Burali-Forti Paradox is named after the Italian mathematician Cesare Burali-Forti. A more modern rendition is presented here. Suppose Buzz Lightyear gives us a mission of the mind, which he calls the Omega Directive, and which will require us to go "to infinity and beyond." The mission is simple: Construct the set of all ordinal numbers.

Seems easy enough. After all, we have just seen how they are ordered, and we have started counting through them, one by one. We have already seen infinite sets, so having an infinite quantity of ordinals shouldn't be a problem. We just conceive of them all in a single set Ω (that's the Greek capital letter *omega*), and it's mission accomplished.

Before we tell Buzz that we've completed the Omega Directive, we pause to recall the definition of an ordinal as the *well-ordered set of all preceding ordinals*. In this scenario, Ω is a well-ordered set of ordinals, meaning that Ω itself is an ordinal, greater than any in its own set. But it should have been in the set if the set contains all ordinals. Contradiction.

We failed in our seemingly achievable mission to construct the set of all ordinals.

The paradox here is *not* that there is no greatest ordinal. Certainly, for any ordinal, there is always a a greater one. There is no greatest natural number, either, but the natural numbers can form a set. The ordinals cannot. Therein lies the paradox.

Claim 1

Infinite sets are possible, so it must be possible to construct the set of all ordinals. We just haven't done it right.

Claim 2

The "set of all ordinals" is an illegitimate concept. The collection of ordinals can be formed only as a proper class.

Discussion and Resolution

Claim 2 is correct. In the words of Bertrand Russell:

> Burali-Forti's contradiction shows that "all ordinals" is an illegitimate notion; for if not, all ordinals in order of magnitude form a well-ordered series, which must have an ordinal number greater than all ordinals.

More generally, recall Russell's warning: "If, provided a certain collection had a total, it would have members only definable in terms of that total, then the said collection has no total." There is no greatest ordinal, and neither is there a set of all ordinal numbers. The ordinals form a class and not a set—a proper class. Advanced readers are encouraged to reread Cantor's Paradox in chapter 1 and look for a connection between the proof that there is no greatest ordinal and the proof that there is no greatest cardinal. This is left as an exercise. Hint: Is the class of cardinal numbers well-ordered?

7.6   Curry's Paradox

Named after the American mathematician and logician Haskell Curry, Curry's Paradox concerns a sentence that declares what its own truth would imply. Consider the following sentence:

> If this statement is true, then aliens built the Pyramids of Giza.

A statement of the form "if this, then that" is called a *conditional* statement. Typically, a conditional claim is proved or disproved by assuming "this" to see if "that" follows. Let's use this proof method to determine whether aliens built the pyramids.

Here, we have a sentence C (for Curry) that says of itself, "If C, then F" (F for falsehood). So we start by assuming our "this," or C, to be true. But C is also the whole sentence, so we have also assumed "if C, then F." Because we have assumed C, F is implied. We have now in fact proved that "if C, then F," which means that C is itself true, which implies F to be true. Hence, the Pyramids of Giza were built by aliens.

Substituting other falsehoods for F, you can prove any crazy thing you wish. That the Earth is riding on the back of a giant tortoise. That the Empire State Building is in Morocco *and* in Zaire, and that ants are little demons from Hades that give birth to hippos. The deduction rules used to prove absurdity were apparently valid. Furthermore, we never had to invoke a concept of negation (calling something false, or denying set membership) as we did in the families of the Liar and Russell Paradoxes. Could it be that everything is true?

Claim 1

It must be that aliens built the pyramids. In fact, Curry's Paradox proves that everything is true. There must be some deeper phenomenon in physics—such as, perhaps, the existence of other universes—that is consistent with the idea that anything and everything is true, even contradictions.

Claim 2

Not everything is true. There are no contradictions. The application of deductive rules used to prove absurdity may be sound in some sense, but there is something invalid about the construction of the proposition itself.

Discussion and Resolution

There are many approaches to resolving Curry's Paradox, and there is no consensus favoring a single resolution. Here, we will apply some concepts already introduced in this chapter. These do not constitute an exhaustive list of approaches.

One approach is hierarchical. The Curry statement refers to its own truth and therefore cannot be expressed in Tarski's hierarchy of languages; a higher-level metalanguage is required to speak of the truth of other statements. Russell's type theory would argue in the same spirit. Kripke would call the statement ungrounded.

Another approach is to express Curry's statement in the language of set theory, beginning naïvely by defining a set X this way:

$$X \equiv \{x | x \in x \to Y\}$$

In other words, X is the set of all sets whose self-membership implies that some statement Y is true. From this definition, we can deduce that X is a member of itself only if its self-membership implies Y:

$$X \in X \to (X \in X \to Y)$$

That is the set-constructed analog of "If this statement is true, then Y is true." Let's ask whether X is a member of itself. If it is, then it follows from what is given here that $X \in X \to Y$. This in turn would mean that X is indeed a member of itself because the previous sentence is itself the condition for its self-membership. Now we have both $X \in X$ and $(X \in X \to Y)$, so Y is also true, whatever it says.

As for resolution, recall that the axiom of specification in Zermelo-Fraenkel set theory prevents us from building sets by accepting "all things that." ZFC requires that we build only subsets from existing sets that are known to be legal. We are allowed to say, "all members of an existing set that." Curry's statement cannot be expressed in ZFC, or in other formal axiomatic set theories.

7.7 Gödel's Incompleteness Theorems

The Austrian-born mathematician and logician Kurt Gödel was no doubt one of the most creative intellectuals of the twentieth century. Gödel emigrated to the United States as a young man. He published his two most famous discoveries—which challenged the very foundation of a mathematical system developed by the giants Bertrand Russell and Alfred North Whitehead—when he was twenty-five years old.

In 1911, Russell and Whitehead sought to provide a comprehensive foundation to all of mathematics with their voluminous magnum opus *Principia Mathematica*. David Hilbert and other mathematicians had worried that the standard axioms of math were inconsistent—that by weaving the axioms together using purely logical steps, one could prove some theorem to be both true and false. Russell and Whitehead took on Hilbert's challenge to lay a foundation of mathematics using pure logic and sets. They wanted to replace verbal language, which might contain subtle ambiguities, with a purely symbolic language that would clarify every nuance.

They set out to create a *metamathematics* that would ascribe meaning to a formal set of symbols (the "language") and a system of rules (the "grammar") about what could be done with those symbols. If everything worked according to plan, by using the terms and grammar of their language, any mathematical truth could be proved, and nothing could be proved to be both true and false. In fact, it would be easy to generate new true formulas simply by creating expressions in their new language using their legal grammar.

But lurking in the shadows was a vulnerability. The system had a security breach. Recall the definition of logic we are using as, in Rand's words, "the art of non-contradictory identification." Formal logic turns propositions into expressions or formulas and evaluates them as true or false, right or wrong, 1 or zero. There is no middle ground.

This posed a challenge for statements that referred to themselves, like the Classical Liar's Statement: "This statement is false."

No worries, thought Russell and Whitehead. It was problematic, they believed, to allow a proposition about a collection of propositions that simultaneously derived meaning from the collection and inserted itself into the collection. They addressed the problem by sorting propositions into hierarchies, as Russell proposed with his theory of types. Any proposition about a collection, they said, would have to belong to a different, higher-level collection. This would help them avoid what they called "vicious circles." With this solution to that pesky security breach of self-reference, they believed they had achieved a logical foundation for mathematics that was complete and consistent. No contradictions allowed.

Then enter Gödel, who discovered that while that security breach had been patched, there was a greater vulnerability at work. Gödel devised a creative way of assigning a unique natural number, called a *Gödel number*, to any symbol, formula, or proof of a formula. He realized that Russell and Whitehead's metamathematics used numbers to prove a maxim about numbers. The subject was being used as a tool to prove something about itself, and that high-level self-reference could be a recipe for inconsistency.

When Gödel took a deep dive in his early twenties, he discovered something remarkable: a true proposition that could be stated using the language and grammar of *Principia Mathematica*, but that was unprovable using the same language and grammar. The statement amounted to, "This proposition is unprovable using the language and grammar of *Principia Mathematica*." If it were provable, then the statement itself would be false, implying Russell and Whitehead's system was inconsistent. But if it were indeed unprovable, then it would be true, but Russell and Whitehead's system could never show it to be true. The system would not be complete.

Claim 1

Common sense tells us that any axiomatic system with basic arithmetic can support a proof of any true statement within that arithmetic, and that when all provable statements are listed, there will be no contradictions among them. Any solid foundation of mathematics is complete and consistent.

Claim 2

Every axiomatic system containing basic arithmetic is either incomplete or inconsistent.

Discussion and Resolution

Claim 2 is correct. Ultimately, Gödel showed that no set of axioms (capable of basic arithmetic) for all mathematics can be both complete and consistent. The result gave rise to the moniker *Gödel's Incompleteness Theorems*.

Any consistent system of axioms would contain true theorems that could not be proved from those axioms. Adding new axioms to patch up the holes would only create new holes elsewhere. Russell and Whitehead's magnum opus, in fact, could never achieve its objective.

This section presents a rough outline of the proof, beginning with the concept of formal systems.

Formal Systems

Recall that a *formal system* is a system that serves as the foundation for deriving logical and mathematical concepts. There are four basic components to a formal system:

- A finite alphabet of *symbols* that can be arranged into strings, creating formulas
- A *grammar* that specifies how a "well-formed formula" is constructed from the alphabet of symbols
- A set of *axioms*, which are themselves well formed
- A set of *inference rules*

In his Pulitzer Prize–winning book *Gödel Escher Bach*, Douglas Hofstadter invents a basic formal system to illustrate the mechanics. He calls it the *MIU-system*.

Hofstadter's MIU-System

The alphabet in the MIU-system consists of the letters **M**, **I**, and **U**. The grammar of the system allows any string that can be constructed using only these letters. All these strings, for example, are in the MIU-system:

> **U**
> **IMUU**
> **MUUUUIIIMMMM**
> **MUIMUIUUMMMIIIUUU**

Although these are grammatically permissible strings, they are not necessarily attainable by the logic of the system. To determine if they are, we need the axioms and rules of inference.

There is one axiom in this formal system, and that is **MI**. There are four inference rules:

- *Rule 1* says that if you have a string ending in **I**, you can append a **U** to the end of the string.

- *Rule 2* says that if you have a string **M***x*, where *x* is everything that follows after the initial letter **M**, then you can generate **M***xx*.
- *Rule 3* says that if **III** occurs in a string, it can be replaced with **U**.
- *Rule 4* says that if **UU** appears in a string, it can be dropped.

These rules are illustrated in the following table:

Rule Number	Rule Statement	Example
1	$x\text{I} \rightarrow x\text{IU}$	**MI → MIU**
2	$\text{M}x \rightarrow \text{M}xx$	**MIU → MIUIU**
3	$x\text{III}y \rightarrow x\text{U}y$	**UMIIIMU → UMUMU**
4	$x\text{UU}y \rightarrow xy$	**MUUUIII → MUIII**

Any string that can be produced, or *derived*, from the axiom and inference rules is called a *theorem*. Hofstadter uses the following illustration of a *derivation* within his MIU-system. Suppose the goal is to derive the theorem **MUIIU**. This can be accomplished by starting with the axiom and applying the rules of inference:

(1)	**MI**	the axiom
(2)	**MII**	from (1) by rule 2
(3)	**MIIII**	from (2) by rule 2
(4)	**MIIIIU**	from (3) by rule 1
(5)	**MUIU**	from (4) by rule 3
(6)	**MUIUUIU**	from (5) by rule 2
(7)	**MUIIU**	from (6) by rule 4

Not every conceivable string of letters **M**, **I**, and **U** is a theorem in the MIU-system, even if the string begins with **M**. It can be proved that **MU** is not a theorem in this system; there is no way to generate it using the rules of inference and the axiom **MI** as a starting point.

A formal system is said to be *complete* if, for any well-formed formula expressible in the system's language, either the formula or its negation is derivable. Such a system is said to be *consistent* if all theorems are compatible; that is, it is impossible to derive a formula and its negation. The MIU-system has no concept of negation, and its symbols have no meaningful interpretation, but the concepts of consistency and completeness do apply to more complex formal systems.

Arithmetization of Formal Systems

One of Gödel's key insights was the idea that the mechanics of a formal system can be embedded into number theory. One can map the typographical rules of a system to a set of arithmetic rules. Hofstadter illustrates this concept using his MIU-system. An example will be helpful here. He begins by mapping each symbol in the MIU-system to a new numerical symbol:

$$\mathbf{M} \leftrightarrow 3$$

$$\mathbf{I} \leftrightarrow 1$$

$$\mathbf{U} \leftrightarrow 0$$

Hence, a string like **MIU** maps to 310. Any numbers would work; Hofstadter chooses these numbers because they look like the corresponding letters. Now, we can rewrite the four rules using the new symbols. For example, the first rule is

> *Typographical Rule 1:* From any theorem that ends in a 1, you can make a new theorem by appending a 0 to the right of that 1.

Notice that this rule can be expressed in terms of pure arithmetic:

> *Arithmetic Rule 1:* From any number $10m + 1$, where m is a natural number, you can make a new number $10 \times (10m + 1)$.

The other rules are more complicated to arithmetize, but their arithmetizations are included here for the sake of completeness. Here, m and k are natural numbers, and n is any natural number less than 10^m:

Rule	If you have...	Then you can create...
1	$10m + 1$	$10 \times (10m + 1)$
2	$3 \times 10^m + n$	$10^m \times (3 \times 10^m + n) + n$
3	$k \times 10^{m+3} + 111 \times 10^m + n$	$k \times 10^{m+1} + n$
4	$k \times 10^{m+2} + n$	$k \times 10^m + n$

Hofstadter has shown an isomorphism between the typographical rules of his MIU-system and the arithmetic rules of his 310-system. The typographical rules of the MIU-system, which describe how symbols can be shifted, changed, dropped, and inserted, have perfectly valid arithmetic counterparts in the 310-system that use addition, subtraction, multiplication, and exponentiation. In the words of Hofstadter:

> This simple observation is at the heart of Gödel's method, and it will have an absolutely shattering effect. It tells us that once we have a Gödel-numbering for any formal system, we can straightaway form a set of arithmetic rules which complete the Gödel isomorphism. The upshot is that we can transfer the study of any formal system—in fact the study of *all* formal systems—into number theory.

Up next is Gödel's clever system of embedding formal systems into number theory.

Gödel Numbering

Gödel's proof relies on a system that can encode any symbol or formula as a unique natural number. There are infinite ways of doing this. The one provided by Gödel works as follows. First, assign to each symbol a natural number. For example:

Symbol	Number
0	1
S	2
+	3
−	4
×	5
÷	6
=	7
∧	8
∨	9
→	10
¬	11
∀	12
∃	13
(	14
)	15

Here, the S symbol is the successor function, meaning "Take the next natural number." The successor to 0 is 1, so

$$S0 = 1$$

Any number n can be represented by invoking the successor function n times on 0. For example:

$$SSS0 = 3$$

The next part, a method for encoding formulas, is quite clever. Suppose you have some formula, like this:

$$0 + 1 = 1$$

First, let's write this out using the successor function:

$$0 + S0 = S0$$

Now we can create an ordered list of the symbols required:

$$0, +, S, 0, =, S, 0$$

Substitute the number for each symbol:

$$1, 3, 2, 1, 7, 2, 1$$

This ordered list of numbers has a length k. Now you multiply the first k prime numbers, raising each nth prime to the power of the nth number on the ordered list. In this case:

$$2^1 \times 3^3 \times 5^2 \times 7^1 \times 11^7 \times 13^2 \times 17^1$$

The fundamental theorem of arithmetic, also called the *unique factorization theorem*, states that *every integer greater than 1 is either prime or the product of a unique combination of primes.* This means that by this encoding method, every formula represented by a string of symbols will map to a *unique* natural number. Incidentally, the formula $0 + 1 = 1$ maps to

$$529, 073, 769, 574, 350$$

This number is so large that it is in the trillions. But it is the unique identifier of the formula at hand. Given only this number, we can work backward and prime-factorize to obtain $0 + 1 = 1$.

We can generalize what we did here to show how to generate a unique Gödel number for any symbol, formula (string of symbols), or proof (list of formulas.) Let σ be the nth symbol on our list of all symbols in the language of a formal system. The Gödel number is given by

$$g(\sigma) = n$$

The function can be extended to output a single number for a string of symbols representing a formula:

$$g(\sigma_1 \sigma_2 \ldots \sigma_k) = 2^{g(\sigma_1)} \times 3^{g(\sigma_2)} \times \ldots \times p_k^{g(\sigma_k)}$$

where p_k is the kth prime number. We can extend the function once again to output a single number for lists of formulas, which constitute proofs:

$$g(X_1 X_2 \ldots X_k) = 2^{g(X_1)} \times 3^{g(X_2)} \times \ldots \times p_k^{g(X_k)}$$

where again p_k is the kth prime number.

Of course, we want to be able to make statements of greater complexity than $0 + 1 = 1$. We need a way to encode variables, just as we would any other symbol. And variables need to be distinguishable by type. For example, we need variables representing numbers, sets, formulas, and proofs, among other types. No problem. Label the different types by the natural numbers. A variable of type n is encoded by p^n, where p is a prime greater than the Gödel number of any basic symbol. (We choose primes here to ensure that each variable gets a unique number.) The following table summarizes this scenario:

Symbol	Number	Type
x_1	17	Numerical variable
y_1	19	Numerical variable
$\vdots$	$\vdots$	$\vdots$
x_2	17^2	Set variable
y_2	19^2	Set variable
$\vdots$	$\vdots$	$\vdots$
x_3	17^3	Propositional variable
y_3	19^3	Propositional variable
$\vdots$	$\vdots$	$\vdots$

Now that we can encode variables, the possibilities are grand. If we started in the appropriate formal system, we could even map a statement of the highly complex Poincaré conjecture to a Gödel number. And since the Poincaré conjecture is the only solved Millennium Problem, we also could map its proof to a Gödel number because we know a proof exists.

Building a Conceptual Ladder

Powerful mathematical tools can be built from simple building blocks. Suppose, for example, that you have a concept of succession (the function S), but no concept of addition. How can the concept of addition be defined from the more primitive concept of succession? It turns out that you can do it this way. Let addition be the function expressed by the symbol +, which takes a pair of natural numbers and maps them to a natural number, and follows these rules:

$$a + 0 = a \qquad \text{(Addition Rule 1)}$$

$$a + Sb = S(a + b) \qquad \text{(Addition Rule 2)}$$

Addition Rule 1 tells us that when you "add" some natural number a and zero, you get the natural number a. Addition Rule 2 tells us that when you "add" a natural number a to the successor of a natural number b, you get the successor to the natural number obtained by "adding" a and b.

This seems very abstract, so let's try applying these rules. Suppose you are new to the concept of addition, so it is not obvious what is obtained by adding $1 + 2$. Whenever you see a + sign, all you know is that the rules must be followed. You are, however, familiar with taking successors, so you can replace a natural number x with S(number before x). Using only your knowledge of successors and the rules of addition as defined, you perform:

$$1 + 2 = 1 + S1 = S(1 + 1) = S(1 + S0) = S(S(1 + 0)) = S(S(1)) = 3$$

Hence, from the concept of succession, we have created a new concept of addition. Let's not stop there. Now that we have defined addition and can comfortably use the + symbol, let's use both succession and addition to define a concept of multiplication. Consider these two rules:

$$a \times 0 = 0 \qquad \text{(Multiplication Rule 1)}$$

$$a \times Sb = a + (a \times b) \qquad \text{(Multiplication Rule 2)}$$

Again, this is very abstract, so let's proceed by example. Let's multiply 2×2, assuming that we are so comfortable with addition that we don't have to break each addition step into a series of successions. Using only succession, addition, and the rules of multiplication as defined, we get

$$2 \times 2 = 2 \times S1 = 2 + (2 \times 1) = 2 + (2 \times S0)$$
$$= 2 + (2 + (2 \times 0)) = 2 + (2 + 0) = 4$$

Once again, our conceptual toolkit has expanded. Starting only with succession, we created a formal concept of addition, and from both follows a concept of multiplication. In the same style, other basic concepts like the inequality relations ($<, >, \leq, \geq$) can be defined. And we can keep building a conceptual ladder from there, so to speak.

These concepts of addition and multiplication have something in common: They are both defined *recursively*. In each case, the second rule uses within itself the very concept being defined. This does not create a problem because the first rule says what happens in the base case, when the addition or multiplication operator has been called within itself so many times that we are left with $a + 0$ or $a \times 0$. Each Rule 1 gives us something real, which can be used to climb out of regress. It's like opening a set of nested Russian dolls, one after another, and there is an innermost doll that contains something other than another doll: an emerald. The formal definition of recursion is omitted here. For our purposes, it's sufficient to understand that concepts that are defined recursively make it easier for us to examine the foundations on which they are built.

Gödel's proof relies critically on a very tall ladder, much like the one we started to build here. In fact, his ladder has forty-six rungs. The ground on which the ladder stands is composed of elementary concepts: basic functions like addition and multiplication, and basic relations like equality and the inequality relations.

The first rung of the ladder is a divisibility relation for variables x and y, which says "x is divisible by y." This is defined like this:

$$div(x, y) \equiv (\exists z)[z \leq x \land x = y \times z]$$

In other words, x is divisible by y if there is a quotient z, where that quotient is less than or equal to x, and y times that quotient equals x. Notice that by its construction, the divisibility relation can be arithmetized via Gödel numbering. This will also be true of concepts higher on the conceptual ladder.

The second rung of the ladder is the prime function for an x. It states "x is a prime number" and is defined this way:

$$prime(x) \equiv \neg(\exists z)[z \leq x \land z \neq 1 \land z \neq x \land div(x, z)] \land x > 1$$

This says that for a number x to be prime, it has to be greater than 1. Additionally, x is not divisible by any numbers less than or equal to itself, other than by 1 and itself.

The prime number function relied on the definition of divisibility. To reach the second rung on the conceptual ladder, the first step had to be reached. So the rungs continue, each building from previous rungs, each attaining a new level of complexity—but each able to be encoded and arithmetized, and also expressed (albeit in ways that are hard to follow) in terms of very simple building blocks. Among the other rungs are the factorial function; a function used to identify the nth prime number by order of magnitude; functions that identify objects as constants, variables, terms, and formulas; and even rungs that identify logical implication. The ladder climbs very high indeed.

The penultimate rung, rung forty-five, formulates a particularly important concept. It introduces a *proof* (x, y) relation, which says in plain English: "x is the Gödel number of a proof of a formula whose Gödel number is y." The previous sentence is unwieldy, but it's critical to understanding what follows. Take a few moments to read and reread it. And remember that because each rung of the ladder is perfectly arithmetized, supported at the bottom and every step of the way by basic primitives, the statement *proof* (x, y) has a Gödel number itself for Gödel numbers x and y.

The highest rung of the ladder, rung forty-six, is the provability function, *provable*(x), which says "x is (the Gödel number of) a provable formula." It is constructed from its conceptual predecessor like this:

$$provable(x) \equiv (\exists y)proof(y, x)$$

In other words, the formula (whose Gödel number is) x is *provable* means that there exists some proof of it (whose Gödel number is) y.

The Shattering Statement

At long last, we have reached the crux of the proof. The property that some statement is provable—that it is a theorem in the formal system—can map to a purely arithmetical property of its Gödel number. That is a remarkable insight, and one that Gödel uses in an ingenious way. Using a method of diagonalization like Cantor's, Gödel constructs a sentence that translates into ordinary English as follows:

This sentence is not provable within the current formal system.

In what follows, let P be the name of this Gödel statement, which says of itself that P is not provable:

$$P \equiv P \text{ is not provable.}$$

Try now to evaluate the provability of P. Is P provable? Let's examine each of the two alternatives: Either P is provable ($\neg P$), or it is not (P).

First let's check the claim $\neg P$, which says that P is provable. If $\neg P$ is a theorem in the system, so is P. A contradiction. If $\neg P$ is a theorem, the formal system is inconsistent. We don't accept contradictions, so we can reject $\neg P$.

The other alternative is P, which says of itself that P is *not* provable. This means that P is not a theorem. And from before, neither is $\neg P$. Everything here is consistent. There are no contradictions. But we have neither P nor $\neg P$ as a theorem. The system is incomplete. Although Gödel uses no notion of truth, you can think of P as an unprovable truth.

These two alternatives boil down to this: In a formal system with basic arithmetic, you must either accept the Gödel statement P and its negation $\neg P$ (inconsistency), or neither P nor $\neg P$ (incompleteness). Within the system, either there is a contradiction or an unprovable "truth." This is the essence of Gödel's Incompleteness Theorems.

Concluding Remarks

There are a few common responses to Gödel's Incompleteness Theorems.

Can't we just add P as an axiom to our formal system? There are no contradictions, so of course we must reject $\neg P$ and accept P. Because P is true, it seems reasonable to add P to our list of axioms. When Henry said, "There's a hole in my bucket, dear Liza, dear Liza," Liza replied: "Well, fix it, dear Henry, dear Henry, dear Henry ..." The problem is, when we patch the hole by adding P as an axiom, we create an entirely new formal system, within which we can construct a new sentence P':

This sentence is not provable within this new formal system.

Trying to patch the hole by adding this as an axiom creates a new formal system, with a new Gödel statement declaring its own unprovability in the new system. Any such patch creates a new hole.

But Russell doesn't allow this kind of self-reference, and neither does Zermelo-Fraenkel set theory! The Gödel statement can be constructed from purely recursive functions and encoded as an arithmetically manipulable Gödel number, within formal systems like that of *Principia Mathematica* and ZFC. Gödel's Incompleteness Theorems apply to every formal system containing basic arithmetic. Gödel's form of self-reference transcends the forms which Russell, ZFC, and other axiomatic set theories strove to patch.

Doesn't the inconsistency implied by ¬P actually prove that P is true? Yes, it does, but only in a way that is outside the original system. This is, in fact, a creative way that mathematicians might be able to prove some challenging mathematical truths. Some have speculated that the Riemann Hypothesis, one of the Millennium Problems, eventually may be proved outside a system if unprovable within a system.

The problematic Gödel statement is artificial and mathematically insignificant; who cares if that's the only statement that is unprovable? The Gödel statement shows the possibility of unprovable truths within a formal system, and its unprovability may not be unique. Mathematicians speculate that there may be other more interesting theorems in ZFC, the standard mathematics, that are both true and unprovable.

7.8   Unexpected Hanging

The Unexpected Hanging Paradox was introduced by D. J. O'Connor in *Pragmatic Paradoxes* in 1948. As the story goes, a judge promises a prisoner: "You will be hanged on a forthcoming weekday, and moreover, the hanging will come as a surprise to you on the day it occurs." The prisoner reasons that the hanging cannot take place on Friday, because if by the end of Thursday the hanging has not occurred, the only remaining day is Friday, but then it would no longer be a surprise. Having eliminated Friday as a possibility, the prisoner reasons that the hanging cannot take place on Thursday, because if by the end of Wednesday the hanging has not occurred, the only remaining day is Thursday, but then Thursday would not be a surprise, either. By an iterative process of backward induction, the prisoner eliminates Wednesday, Tuesday, and Monday as possibilities, then says to the judge: "You cannot deliver on your promise." But then a day comes—say, Wednesday—when the judge shows up at the prisoner's cell, ready to hang him. This comes as a surprise to the prisoner. The judge's promise has come true after all. What was wrong with the prisoner's reasoning?

(A less morose version of the paradox tells the story of a teacher promising students that a surprise exam will take place during a coming weekday.)

Claim 1

If we define terms like *surprise* rigorously, we can show that the judge's pronouncement is self-contradictory. That the hanging took place, and that it came as a surprise, do not change the contradictory nature of the judge's pronouncement.

Claim 2

The judge implies that the prisoner can know that he is unable to know something. We have to be careful about how we treat knowledge in this scenario. We must define *knowledge*, along with a standard of knowing that is consistent.

Discussion and Resolution

This is a tough paradox to crack and categorize. So far, it may seem this paradox belongs in chapter 8, on induction, seeing as the prisoner uses backward induction to conclude the judge is lying. You will soon see why this was nonetheless included in the current chapter, and depending on its treatment, can be called an "epistemic paradox," or paradox of knowledge.

Academics have not reached a consensus on the resolution to this paradox. The two claims given here describe the essence of the two main approaches to the problem: respectively, the *logical* and the *epistemological* approaches. Both approaches are enlightening and complex, so we'll take a dive into each. The deepest possible dive would require learning entire systems of notation. For the purposes of this book, we'll dive deeply enough to need scuba gear, but we'll avoid submersibles.

Logical School

Recall from the section on Gödel's Incompleteness Theorems earlier in this chapter that Gödel established a system by which to logically express provability. Symbols, formulas, and proofs can be mapped to natural numbers in decipherable ways. The system can be used even to represent provability itself.

If we are going to express a proposition logically, we first need to have a clear understanding of its meaning. Our task is to choose a sensible interpretation of the judge's pronouncement that we can formalize. Our interpretation also needs to support the backward inductive argument used by the prisoner. Articles by mathematicians Timothy Chow and F. A. Fitch help us with the task of defining *surprise* through a process of iterative refinement. We'll begin by revisiting the judge's proposition:

> You will be hanged on a forthcoming weekday, and moreover, the hanging will come as a surprise to you on the day it occurs. (a)

Now let's rephrase the proposition in a way that attempts to formalize the definition of *surprise:*

> You will be hanged during a forthcoming weekday, and moreover, the date of the hanging will not be provable in advance from an empty set of axioms. (b)

The *from an empty set of axioms* clause may seem like a useless appendage, and in fact it is. It was included here to highlight its own uselessness. If we are going to speak about the provability of the date of the hanging in advance, we must declare a framework of axioms from which to establish provability. A statement cannot be provable without the tools to prove it. We need to specify at least one such tool. Let's try again:

> You will be hanged during a forthcoming weekday, and moreover, the date of the hanging will not be provable in advance given that you will be hanged on a forthcoming weekday. (c)

This phrasing is more attractive because it ties provability to an axiom; it offers a tool and says that tool cannot be used to prove the date of the hanging. However, we are going to need a better tool.

Using this proposition alone, the prisoner's logic can be used to eliminate Friday. At the end of Thursday, Friday would be provable, given that the prisoner will be hanged on a weekday. However, progress from here is limited. At the end of Wednesday, this proposition is not enough to eliminate Thursday. The logic cannot continue backward because provability depends only on the axiom that "the prisoner will be hanged on a forthcoming weekday," not on the whole proposition itself. The proposition needs to acknowledge that the date of the hanging cannot be proved from itself. Okay, on to the next iteration:

> You will be hanged on a forthcoming weekday, and moreover, the date of the hanging will not be provable in advance given this proposition. (d)

Now the proposition takes itself as the axiom. This will allow the sort of backward induction that the prisoner uses in the original statement of the paradox. The next step is to express this statement in formal, logical language and determine whether it holds water or contradicts itself. To simplify things, let's reduce the week to two days, Monday and Tuesday. Assume that these are the only weekdays to which the judge is referring. First, some notation:

Symbol	Meaning
Q_1	the statement "the hanging takes place Monday"
Q_2	the statement "the hanging takes place Tuesday"
or else	this or that, but not both
$P(X)$	"Statement X is provable"

Now let's translate (d) into a logical statement S:

$$S \equiv [Q_1 \wedge \neg P(S \rightarrow Q_1)] \text{ or else } [Q_2 \wedge \neg P(\neg Q_1 \wedge S \rightarrow Q_2)] \qquad (7.1)$$

In English, this statement saying that either (1) the hanging takes place on Monday, and the fact that it takes place on Monday cannot be proved from this statement; or (2) the hanging takes place on Tuesday, and this statement—in conjunction with the nonoccurrence of a Monday hanging—does not prove that the hanging will take place Tuesday.

Using equation 7.1 as a starting point, F. A. Fitch proves in a Gödelian fashion that S is self-contradictory. The proof is omitted (in order to keep the promise of no submersibles). Now, we have yet to interpret Fitch's proof of self-contradiction. To what extent does it resolve the paradox, given that the judge's promise appears to be vindicated after the fact?

To answer that, we need to know if the judge's promise was really vindicated. Fitch shows us that judge's pronouncement (d) is provably false. It is logically inconsistent and nonsensical, and therefore an invalid source, cause, or definition of surprise. The judge's "surprise" appearance on Wednesday must qualify as such under a different definition, which could be provided by interpretation (c) of the pronouncement.

Under interpretation (c), and using the Laplace principle of indifference introduced in chapter 4, on probability, the prisoner assigns uniform probability of a hanging on all days from his current day through Friday. So at the start of Monday, the hanging could take place Monday through Friday, with probability $\frac{1}{5}$ on each day. At the start of Tuesday, the hanging could take place Tuesday through Friday, with probability $\frac{1}{4}$ on each day.

The logic continues for Wednesday, Thursday, and Friday, with respective probabilities $\frac{1}{3}$, $\frac{1}{2}$, and 1 on each remaining day.

We can now recast the original story in a way that encapsulates the reasoning given here:

The Complete Logical Story

A judge promises a prisoner: "You will be hanged on a forthcoming weekday, and moreover, the hanging will come as a surprise to you on the day it occurs." First, the prisoner interprets and logically formalizes the judge's pronouncement as (d), but following a Gödelian proof, he realizes this interpretation to be nonsensical. Next, he interprets it as (c). Using the rules of conditional probability and Laplace's principle of indifference, at the start of each day he reaches, he assigns uniform probability of a hanging on each remaining possible day. On each of days Monday through Thursday, that probability is less than 1 and thus "surprising" under the definition in (c). Hence, the judge's promise is vindicated for any hanging on Monday through Thursday. The judge arrives on (say) Wednesday, and the prisoner is "surprised"—but not surprised that he is surprised.

This is a self-consistent story, and a resolution to the paradox.

Some have argued that the prisoner's backward inductive logic defeats itself. The self-defeatists argue that the prisoner rationalizes away the possibility of a surprise, and in so doing, opens up the very possibility of a surprise again. If the prisoner has deluded himself into thinking there can be no surprise, then he can be surprised; but if he believes he can be surprised, he can rationalize away the possibility of surprise. The self-defeatist interpretation of the paradox is akin to the Liar's Paradox: "This statement is false." It creates a sort of vicious cycle, they say, with no stable conclusion.

But they don't have it quite right; their resolution is too simplistic. The Logical School provides a resolution that is more nuanced than that. When the prisoner rationalizes away the possibility of a surprise, he does so according to a specific definition of *surprise*—say, in (d), as provided by the Logical School. The prisoner's conclusion is not that a surprise is impossible by any definition, or that he won't be hanged. The prisoner's conclusion is that the judge's pronouncement is provably self-contradictory. Recognizing the inconsistency of interpretation (d), the prisoner can then choose a different interpretation of the judge's promise, and one that does not preclude a surprise hanging within a new definition.

Epistemological School

The epistemological school puts the prisoner's knowledge and beliefs under a microscope. How does the prisoner react to the judge's statement? Does he believe it? Trust it? Know it? Choose a more suitable interpretation? Do his beliefs and knowledge stay with him through the week? The fundamental questions of the epistemic approach are: What is knowledge? How does one gain and keep it?

Much has been written about the Unexpected Hanging Paradox from the epistemological perspective. Some scholars have formalized systems of epistemic logic to address it. This section offers just a taste of this approach.

For simplicity, let's continue with the two-day scenario. The hanging can take place either on Monday or Tuesday. Let's interpret the judge's pronouncement in the following way:

> Either (1) the hanging will be on Monday, and you will not know this
> before it occurs on Monday, or (2) the hanging will be on Tuesday, and   (e)
> you will not know this before it occurs on Tuesday.

Suppose the prisoner survives Monday. He knows that he can eliminate the first half of the pronouncement because the hanging did not take place on Monday. He is left with a proposition that boils down to: "The hanging will take place today, but you don't know that."

This perplexing statement is a version of Moore's Paradox, which asserts the truth of a proposition in conjunction with a statement that the proposition is not known. A common example is, "It is raining, but I do not know that it is raining." Subtle variations of Moore's Paradox assert knowledge versus belief, present or future tense, or first or second person. These statements are considered paradoxical because, despite their absurdity, they can be true and logically consistent.

Moore's Paradox is similar to the self-referential Knower Paradox popularized and named by David Kaplan and Richard Montague, in which a statement declares its own unknowability. Such a declaration is problematic under two assumptions. The first assumption is that knowledge is truthful; if P is known, then P is true. The second implies that proven truths constitute knowledge; if P is proved, then P is known. Now let's consider some statement K, defined as, "This statement is not known." If K is known, then it is true, but its truth implies that it is not known. But if it is not known, then it fulfills its own truth, and provably so, which makes it known—a contradiction.

Moore's Paradox creates a phenomenon in epistemic logic known as a *blind spot*, a proposition that is consistent but inaccesible at certain points in time to certain people. The proposition faced by the prisoner—"the hanging will take place today, but you don't know that"—is a blind spot to the prisoner. It is a statement that he cannot know. Anyone else can know it, but the prisoner cannot.

To see why the prisoner cannot know it, let's test the veracity of a general statement, "*P*, and *P* is unknown." Notationally, if we know something, we put a *K* before it. Suppose someone claims it is known that "*P*, and *P* is unknown." Let's examine the statement logically; see the following table (The proposition *P* will be written in lowercase so knowledge is more distinguishable.)

Step	Logical basis
1. $K(p \land \neg Kp)$	By assumption
2. $Kp \land K(\neg Kp)$	By distributing knowledge over 1
3. $\neg Kp$	By the truthfulness of knowledge in 2
4. $Kp \land \neg Kp$	By 2 and 3, a contradiction
5. $\neg K(p \land \neg Kp)$	By 1 and 4, reductio ad absurdum

We have demonstrated that within a reasonable system of assumption, it cannot be known that "*P*, and *P* is unknown." The latter is a blind spot, even if it still can be true.

If we restrict whom we're talking about, the blind spot can be avoided by some people. For example, Nick can know that "*P*, and Travis does not know *P*." But Travis cannot dodge the blind spot, just as the prisoner cannot dodge the blind spot when he faces: "The hanging will take place today, but you don't know that."

What's remarkable is that the prisoner has not forgotten anything, but he has been forced to unknow some of the content of the original announcement. Some adherents to the epistemic school say that this is enough to resolve the paradox. The hanging, they say, can still take place on the last day and it can still surprise the prisoner because by the time the last day comes around, the prisoner is forced to unknow a certain truth. The very fact of unknowing makes it truthful. The prisoner may think he can eliminate Friday, and then Thursday, and so on, but this reasoning is flawed. He can't actually eliminate Friday, because on Friday he will be forced to unknow the content of the judge's promise.

   The literature on this approach is vast. This section has offered but a taste of the epistemological approach to the Unexpected Hanging Paradox. The general idea is that certain reasonable assumptions about knowledge are inconsistent with the prisoner's assumption that he can sustain his own knowledge of the judge's proposition.

8 Induction

Induction … is a proceeding from particulars to a universal.
—Aristotle

It is not immediately clear what Aristotle meant by "proceeding from particulars to a universal," but there are two possible interpretations. One interpretation is movement from particular propositions, like "Here is a white swan," and "There is a white swan"—to universal propositions, like "All swans are white." The second interpretation is movement from particular units with similar characteristics—"Here is a white bird with a long neck," and "There is a white bird with a long neck"—to the universal concept of *swan*.

John McCaskey, a scholar of the history of science, calls the history of induction "a back-and-forth between these views," beginning with the ancient thinkers:

> Socrates, Aristotle, Cicero, and basically everyone else in the ancient world thought induction was the second. It was that thing Socrates did when he asked, for example, "What is piety?" He identified some instances and then, using an iterative process of comparing and contrasting, identified the essence, identified what made piety piety, identified what Aristotle called the formal cause—or at least Socrates tried to. He never seemed to get there, and we can't tell for sure if he thought you could. But Aristotle thought you could. His *Posterior Analytics* and especially parts of his *Topics* were guidebooks on how to do so, how to work up from observation of particulars to conceptions of universals.

The pendulum would start to swing back to the first view, McCaskey says, with the Neoplatonists of late antiquity, and stay with the Scholastics through the Middle Ages. It would swing again to the second during the Renaissance, largely with the help of Francis Bacon, and in the early 1800s, it would swing back to the first, where it has stayed up to now.

Because induction is now most associated with the first interpretation, it is also closely associated with philosophical scandal and logical fallacy. The so-called problem of induction is the idea that one cannot generalize about the characteristics of an entire class of things after observing only a limited number of instances. If, for example, you have seen hundreds upon hundreds of white swans, you might be tempted to say "All swans are white." But then, what do you do when you see a black one? The instances in the class could also be events. For example, the laws of physics have held every time they have been observed, so we might infer that they will continue to hold—but is such an inference logical?

This first view of induction, the propositional and problematic one, presupposes the need for valid concepts and a process by which to form them. One cannot say, "This swan is white" or "This swan has a long neck" without first having a concept of *swan* and a standard by which it can be known if a certain thing qualifies as a swan. With respect to the laws of physics, one cannot begin to evaluate the uniformity of gravitational force without first having concepts of mass and distance. Nor could one say whether "fools fall in love" without a standard of knowing what is or isn't a fool, and what is or isn't love. Socratic induction, says McCaskey, was meant to provide that standard.

There is another kind of induction, and that is mathematical induction, also called *recurrence*. In some sense, this kind of induction is also a proceeding from particulars to universals. In *The Principles of Mathematics*, Bertrand Russell writes:

> Mathematical induction, which is purely ordinal ... may be stated as follows: A series generated by a one-one relation, and having a first term, is such that any property, belonging to the first term and to the successor of any possessor of the property, belongs to every term of the series.

Let's unpack Russell's description a bit. If mathematical induction is to apply, the first thing we need is an ordered series of things. The series needs to have a first term. Every term needs to have a unique predecessor (except the first term) and successor (except the last term, if there is one). In other words, we need to be able to label the terms in the series by $n = 1, 2, 3$, and so on.

Mathematical induction is a technique used to prove, in two steps, a statement about all the things in the series—typically, that all the things have some property. The first step, called the *base* step, is to prove that the first term, term 1, has that property. The second step, called the *inductive* step, is to prove that if some term n has the property, then term $n+1$ also has the property. If you succeed in each of these two steps, then you have proved that every term in the series has the given property.

To use a slightly more formal language of logic, mathematical induction is used to prove a proposition, or predicate, to be true for every term in a suitable series. A proposition, or predicate, is a *Boolean-valued* function on the terms in the series—that is, a function that maps each term to one of two values, which we interpret as true or false. You can think of the predicate function $P(n)$ as asking whether term n has some property P. The answer to that question is either yes or no. The statement "term n has property P" is either true or false. "True" is often represented by a 1, and "false" by a 0. The predicate function outputs a single bit of information for a given input.

Using this more formal language, mathematical induction is expressed by the following axiom: For any property P, if $P(1) = true$, and if $P(k) = true$ implies that $P(k+1) = true$ for all k, then $P(n) = true$ for all n. Here, k and n are counting numbers and should be thought of as labels for terms in a series. We don't necessarily have to be talking about the numbers (labels) themselves. We could be talking about cows in a line, labeled $1, 2, 3$, and so on. If we are talking about cows, then $P(n) = true$ means that the nth cow in our line has property P.

Are we still proceeding from particulars to universals? We are, in the sense that we must examine a single term in the base step and a generic pair of terms in the inductive step. Together, these steps allow us to make a general statement about the series as a whole, without testing every single term.

We've now covered three different meanings of *induction* in logic. Propositional and Socratic induction are epistemological concepts pertaining to how we learn and reason at a fundamental level. Mathematical induction is a technique used for proving statements about all terms in a series of things. Could there be a fourth?

In fact, there is: backward induction. *Backward induction* is the process of reasoning used to determine optimal actions at different points in time when decisions can be made. The reasoning goes backward in time, starting

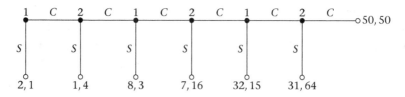

Figure 8.1
A centipede game.

with the ultimate decision to be made, whose optimality helps determine the best penultimate decision, whose optimality helps determine the best decision before that, and so on, until the best course of action can be determined at every point in time and for every possible scenario—back to the beginning.

Backward induction is used extensively in game theory to calculate *equilibria* (sets of optimal actions) in games where players take turns. An example will be helpful here. In 1981, Robert Rosenthal introduced a thought experiment called the centipede game, wherein two players take turns deciding whether to continue or stop the game. The game continues for a limited number of turns. Each time the game continues, the sum of payoffs between the players increases were the game to suddenly stop, favoring the player whose turn it becomes.

Consider the game as drawn in figure 8.1. The players are called Player 1 and Player 2. The decision nodes are marked with solid dots, and labeled according to which player's turn it is to act. The hollow nodes represent the termination of the game. At each hollow node there is a pair of numbers representing the payoffs to Players 1 and 2, respectively. Higher numbers are better; you can think of these as dollar rewards for winding up at that node. Each line segment is labeled by the decision it represents: *S* for stop, or *C* for continue.

There are six decision nodes. We will refer to them from left to right as nodes 1 through 6, even though they are labeled by the player whose turn it is. An equilibrium for this game would tell us each player's optimal behavior at each node; that is, it would specify an action for Player 1 at each odd-numbered node and an action for Player 2 at each even-numbered node. That equilibrium will give us Player 1's optimal move at the start of the game.

Intuitively, it seems like the players should continue the game as long as they can. If they could only cooperate and agree to keep playing, a possible reward of $50 each awaits them—big money compared to the measly $2 and $1 they'd receive if the game were to terminate immediately.

There are a few important assumptions about the players. First, each player only cares about himself, and his only goal is to maximize his own payoff. Both players know this. It doesn't matter if they are friends or enemies. A player derives no happiness or displeasure if the other player's wallet gets fatter. All that matters to a player is the size of his own wallet at the end of the game. So there really is no trust if one player makes a promise to do something in the future that would be suboptimal for him at that point in the future. Both players know this about each other.

One more thing must be achieved before we solve for the equilibrium, and that is to understand what an equilibrium is. An *equilibrium* is a mathematical construct with a specific definition, which is described extensively in chapter 6. An equilibrium specifies a strategy for each player, such that each player's strategy is optimal given the other's strategy. Equilibrium does *not* mean that two real human people would necessarily play this way in real life. People are fickle. They have free will. But free will is irrelevant to the equilibrium. All that matters are strategies that meet specific conditions, given the structure of a game and its incentives.

We are now ready to apply the concept of backward induction to solve for the equilibrium of this game. We start with the final decision node, node 6, where it is Player 2's turn. Player 2 has a choice between the actions Stop and Continue. If he Continues, Player 2 gets 50; if he Stops, he gets 64. So at node 6, Player 2 chooses the action that gives him the greater payoff: He Stops.

At node 5, Player 1 knows that Player 2 will Stop at node 6. So Player 1 chooses between Stop at node 5, for a payoff of 32, or Continue at node 5, which he has figured will yield himself a payoff of 31. Stopping is (slightly) more favorable. At node 5, it is optimal for Player 1 to Stop.

At node 4, Player 2 knows that Player 1 will Stop at node 5. Player 2 must choose between a payoff of 16 (Stop) or 15 (Continue). He Stops.

At node 3, Player 1 knows that Player 2 will Stop at node 4. Player 1 must choose between a payoff of 8 (Stop) or 7 (Continue). He Stops.

At node 2, Player 2 knows that Player 1 will Stop at node 3. Player 2 must choose between a payoff of 4 (Stop) or 3 (Continue). He Stops.

At node 1, Player 1 knows that Player 2 will Stop at node 2. Player 1 must choose between a payoff of 2 (Stop) or 1 (Continue). He Stops.

We have solved for the equilibrium: Each player will stop at any node he reaches. This manifests with Player 1 stopping as soon as the game starts.

This is the process of backward induction. Have we proceeded from particulars to a universal? Sort of. Certainly not in the sense that Aristotle was talking about, and not in the sense of mathematical induction, either. But we have proceeded through a series of particular decision nodes, working backward in time, to solve for equilibrium—a concept that encompasses the entire game and specifies a complete strategy for each player.

8.1 All Horses Are the Same Color

The following paradox was introduced by George Pólya in *Induction and Analogy in Mathematics*. The goal is to prove that all horses are the same color. Now, you may think there's no way you could ever read a convincing argument of such an outlandish proposition. Hold that thought. When given the argument for the first time, even trained mathematicians have been known to find its apparent logic disconcerting at first.

As we proceed, it will be helpful to use an adjective to describe sets of things that are all the same color. The word we will use is *monochromatic*. If a set of things is monochromatic, then all its elements are the same color. For example, the set of all rubies is monochromatic because all rubies are red.

If our goal is to prove that all horses are the same color, we can achieve it by proving that any set of n horses is monochromatic. The latter statement implies the former. Before proceeding, make sure you have stopped to understand why. From here on, it will be our goal to prove that "for any n, a set of n horses is monochromatic." We will prove that statement using mathematical induction.

The first step in mathematical induction is the base step. We must prove our proposition to be true for $n = 1$. This is trivial: Any single horse indeed has a color of its own. In our more formal language, any set of a single horse is monochromatic.

The next step is the inductive step. We must prove that if all sets of n horses are monochromatic for some n, then all sets of $n + 1$ horses must be monochromatic. Because we are trying to prove an "if then" statement, we are allowed to assume the condition to learn whether the condition implies the consequence.

So let's assume for a moment that the condition is true: Any set of n horses is monochromatic for some number n. Now let's suppose that we have a set of n horses before us. The condition implies that our set is monochromatic. They are all the same color. Now let's introduce a new horse to the set. Call our new horse Shadowfax. Including Shadowfax, we now have $n + 1$ horses. Let's temporarily kick out another horse, named Seabiscuit. Our current set of horses, which includes Shadowfax and excludes Seabiscuit, has size n—and therefore, must be monochromatic according to our condition. Therefore, Shadowfax must be the same color as all the others. Now let's bring Seabiscuit back into our set. We know that Seabiscuit was the same color as all the others to begin with. Our current set of $n + 1$ horses, which includes both Shadowfax and Seabiscuit, must be monochromatic.

We have proven that if it is true that all sets of n horses are monochromatic for some n, then it must be true that all sets of $n + 1$ horses are monochromatic. The "if" condition implies that any new horse introduced to a set of n horses must be the same color as all the others.

That is all. The proof is complete. We have successfully completed the base step and the inductive step. We have shown that any set of one horse is monochromatic (trivially), and that if all sets of n horses are monochromatic, then all sets of $n + 1$ horses are monochromatic.

With mathematical induction as our tool, we have shown that all sets of n horses, for any n, are monochromatic. In other words, all horses are the same color. And there is absolutely nothing wrong with mathematical induction.

We don't have to stop there. Color is but one property. We can replace that property with any other property: size, shape, speed, and so on. And we aren't restricted to horses. We can in fact use this same logic to argue that all things have all the same properties. Suddenly, our world starts to look uniform, like the world according to the Parmenidian view (see chapter 2, on Zeno's Paradoxes of Motion). Is difference even possible?

Can you spot the logical error?

Claim 1

Mathematical induction is axiomatically valid. We have applied it correctly here, so the conclusion must be correct. Every horse is in fact the same color as every other horse.

Claim 2

There is something loaded in the word *monochromatic*. It oversimplifies things. After all, single horses can be multicolored. Some of them have spots. The word *monochromatic* unrealistically forces any given horse to be of a single color. The logical error lies in the grammar.

Claim 3

The proof relied on assuming something ridiculous. For the inductive step, we had to assume that "any set of *n* horses is monochromatic for some *n*." We aren't allowed to do that.

Claim 4

Not all horses are the same color. We know that. But mathematical induction proves otherwise. Therefore, there must be a problem with mathematical induction itself.

Claim 5

There is nothing wrong with mathematical induction when properly applied, or the word *monochromatic,* or the assumption we made in the inductive step. The error lies elsewhere—someplace hidden.

Discussion and Resolution

This tricky paradox has fooled many an expert. It is not immediately obvious where the error lies. In fact, the error is so well hidden that Claim 1, however preposterous, becomes somehow tempting. You might have started asking yourself whether you really have seen horses of different colors. But Claim 1 is definitely false.

Claim 2 is as well. Yes, some horses are multicolored, but this fact is irrelevant to the mathematics at play. The word *monochromatic* could be redefined to mean "having the same pattern of colors," and there would be no problem. The property of coloration is not what matters; the proof can be used to show that "all horses have property *X*" just as well. Insert whatever you'd like for *X*.

Claim 3 is also tempting, but false. We are not actually assuming something ridiculous in an improper way. In the inductive step, we did assume that "any set of *n* horses is monochromatic for some *n*," but only to test the implications of that statement—specifically, whether it implies the same condition for any set of $n+1$ horses. Ultimately, we were not making any permanent or improper assumptions. We were only trying to determine whether one thing, however ridiculous it may have sounded, would imply another thing if that first thing were true. We are allowed to do that.

Claim 4 is false. There is nothing wrong with mathematical induction when properly applied.

That takes us to Claim 5, the correct claim (albeit uninformative, which was by design, to prevent spoilers). The error lies in our application of the inductive step. The error is easy to overlook. And in a sense, the inductive step is *almost* valid. It works for any n except $n = 1$.

If it were true that all sets of 2 horses were monochromatic, then indeed we could conclude that all sets of 3 horses were monochromatic. And if all sets of 3 horses were monochromatic, then we could say the same of all sets of 4 horses. And so on. But we *cannot* say "if any set of $n = 1$ horse is monochromatic, then any set of 2 horses is monochromatic."

To see why, suppose we have one horse named Seabiscuit. All single horses are monochromatic; trivially, Seabiscuit is its own color. Now Shadowfax comes along and stands next to Seabiscuit. Shadowfax must be his own color, too. But he does not have to be the color of Seabiscuit. The inductive step does not work for $n = 1$.

It's easy to make a bad assumption about the inductive step: that it matters only for big numbers. The base step handles $n = 1$, a small number, so the inductive step should take care of the bigger numbers. When we read "any set of n horses is monochromatic for some n," we assume that n is large, especially having just handled the base case.

That's faulty thinking. The base step must prove a property to be true for the first term. The inductive step must prove the property belongs to the successor of *any* possessor of the property.

An analogy is helpful here. Suppose there were an infinite staircase, as in figure 8.2. Each step is only a foot below the next, except the first step, which is a mile below the second step. You are currently standing on the first step. Can you climb onward to infinity? Of course not. You can hardly begin. Even though you are standing on the first step, and nearly every future step would be accessible from the previous one, you cannot reach step 2 from step 1. If you are told "for all $n \geq 2$, each step n is only a foot below the next, and you are currently standing on the first step," it is easy to overlook the challenge of reaching step 2.

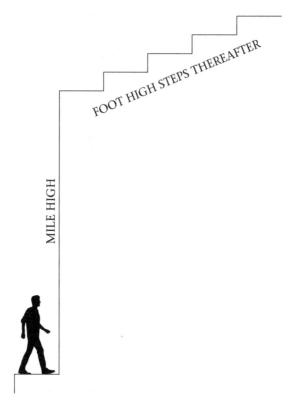

Figure 8.2
The infinite staircase with a mile-high second step (not to scale).

For the same reason, it is easy to miss the fallacy in the horse paradox.

8.2 Blue-Eyed Islanders

A logically equivalent version of the following puzzle was introduced by
Herbert Gintis in 2000, called the "Women of Sevitan." There are three
steps to understanding this paradox. In the first step, a puzzle is introduced.
In the second step, the puzzle is solved. In the third step, a key aspect of
the solution is challenged, and therein lies the paradox. Let's start with the
puzzle.

The Puzzle
A deeply religious tribe resides on an island. There are 1,000 tribesmen: 900
with brown eyes, and 100 with blue eyes.

There are no reflective surfaces on the island (not even the water), so no tribesman can ever see his own eye color, but every tribesman can see every other tribesman and count eye colors accordingly. Speaking of anyone's eye color is forbidden, so a tribesman never learns his eye color verbally from anyone else. Any tribesman who somehow learns his own eye color will commit ritual suicide at noon the next day, in the public square, for all to see. Each tribesman is a perfect logician, meaning that he will make any logically deducible conclusion instantly. Each tribesman knows all the information contained in this paragraph.

For years, the tribe has lived in harmony. No one has committed suicide. One day, an explorer (say, with green eyes) lands on the island, unaware of the tribe's superstition, and makes a pronouncement for all to hear at a gathering: "It's remarkable to see at least one person with blue eyes in this part of the world." What effect does his faux pas have on the island's population?

The Solution

On the 100th day after the announcement, all blue-eyed islanders will commit suicide. A more general statement can be made: If there are k islanders with blue eyes, all of them commit suicide on the kth day.

Suppose $k = 1$, meaning that there is one person on the island with blue eyes. After hearing the explorer's announcement, the only blue-eyed islander—seeing everyone else with brown eyes—deduces that the person with blue eyes must be himself. So he commits suicide the next day at noon. Hence:

Theorem 1 If $k = 1$, the 1 blue-eyed islander commits suicide on day 1.

Now suppose $k = 2$. Each blue-eyed islander sees someone else with blue eyes, and therefore cannot deduce on day 1 that he himself has blue eyes. Neither commits suicide on day 1. But then each one realizes from Theorem 1 that if only one of them had blue eyes, that tribesman would have committed suicide on day 1. Each one then deduces that he himself has blue eyes, and commits suicide on day 2.

Theorem 2 If $k = 2$, the 2 blue-eyed islanders commit suicide on day 2.

The logic continues in a similar fashion up to any k:

Theorem n If $k = n$, the n blue-eyed islanders commit suicide on day n.

The solution can be proved by mathematical induction. Theorem 1 handles the base case. The inductive step must prove that if Theorem n

holds, then Theorem $n+1$ must hold. Let's prove the inductive step. Suppose $k=n+1$. By observational counting, each blue-eyed islander knows that either $k=n$ or $k=n+1$; he does not know which, since he cannot see his own eye color. On day n, no suicides occur because at that time, each blue-eyed islander still does not know if $k=n$ or $k=n+1$ (i.e., does not know his own eye color). But by Theorem n, if $k=n$, there would have been n suicides on day n, which there were not. Each blue-eyed islander therefore concludes $k=n+1$, and that he himself has blue eyes. All $n+1$ of them commit suicide the next day.

We have now shown the base case (Theorem 1) and the inductive case (Theorem n implies Theorem $n+1$), so by mathematical induction, Theorem n is true.

This is a challenging logical puzzle. It's important that you understand the solution presented in this section before continuing. If the inductive proof seems convoluted, start by making sure that you understand Theorem 1. Once that makes sense, proceed to Theorem 2. Before jumping to a general Theorem n, you might try moving from Theorem 2 to the next logical step: Theorem 3. Moving from one scenario to the next will help you understand the logic used to prove the general statement for k blue-eyed islanders.

The Paradox
Now, if the solution weren't difficult enough, here's a paradox within it.

The solution implies that if 100 islanders have blue eyes, they will all commit suicide 100 days after the explorer's pronouncement.

But all the explorer effectively communicated was, "There is at least one person with blue eyes on this island." Every one of the islanders already knew this piece of information. Furthermore, every islander knew that every other islander knew this piece of information, too. Why, then, does the explorer's statement trigger the mass suicide? Shouldn't the suicides have happened long ago? How have the islanders been living in harmony when each has known—and each has known that each has known—of the presence of at least one blue-eyed islander among them for years?

Claim 1
So long as more than one islanders have blue eyes, the explorer's statement indeed conveys no new information. If there are at least two islanders with

blue eyes, each can see that "there is at least one person with blue eyes on the island," and that is all that is necessary to trigger the eventual suicides. Therefore, if $k > 1$, the suicides will have indeed taken place years ago. The explorer's statement matters (provides new information) only if $k = 1$.

Claim 2

The public nature of the explorer's statement matters. While "there is at least one person with blue eyes on this island" is information that all islanders knew before the announcement, the fact that it was announced *publicly* gives rise to a new form of common knowledge, which permits further deduction and instigates the suicides.

Discussion and Resolution

Claim 2 is correct. This puzzle is often used as a case study in common knowledge. Let's start with some new concepts and notation.

Let P be a proposition, and let $E(P)$ be the proposition that "everybody knows P." A statement $E(P)$ constitutes *universal knowledge* because it is something that "everyone knows." One can also assert the universality of universal knowledge by $E(E(P))$, or "Everybody knows that everybody knows P," and denote this as $E^2(P)$. This can scale up ad infinitum. One can also make a stronger assertion that "for every natural number n, it is true that $E^n(P)$" and denote this proposition of *common knowledge* by $C(P)$.

The explorer's public announcement represents a leap from universal knowledge to common knowledge. Before the explorer's knowledge, each of the 100 blue-eyed islanders has the universal knowledge that $E^{99}(k \geq 1)$, but *not* that $E^{100}(k \geq 1)$. The explorer's announcement introduces the common knowledge $C(k \geq 1)$. Specifically, it is the instance $E^{100}(k \geq 1)$ that sets the eventual mass suicide in motion.

To see why, let's look at specific cases with increasing complexity and study the behavior of the islanders before the explorer makes his announcement. The preannouncement conditions will give rise to a new set of theorems.

First, suppose $k = 1$. There is one blue-eyed islander. That islander cannot see his own eye color and has no way of deducing it. Therefore, he will not commit suicide:

Theorem 1 If $k = 1$, no suicides will occur.

Let's proceed to the next level of complexity: $k = 2$. To make it easier, let's name our blue-eyed islanders Travis and Nick. Each can see that the other has blue eyes. Let's take Travis's point of view. Travis knows that $k \geq 1$, but he does not know whether Nick knows that $k \geq 1$, since Travis does not know his own eye color; he doesn't know what Nick sees when looking at him. Nick's inaction after one day does not provide Travis with any information about Travis's eye color. The argument is symmetrical for Nick with respect to Travis. We can say $E(k \geq 1)$, that "everyone knows $k \geq 1$," but it is not true that $E^2(k \geq 1)$. Neither Nick nor Travis can deduce anything about his own eye color; therefore, neither will commit suicide.

Theorem 2 If $k = 2$, no suicides will occur.

Now add a third level of complexity: $k = 3$. The blue-eyed islanders are Travis, Nick, and Jeff. Again, let's use Travis's point of view, which applies symmetrically to the others. Travis (and each other blue-eyed islander) knows that $k \geq 1$. Travis knows that Nick knows that $k \geq 1$, since Travis knows that Nick can see Jeff. And Travis knows that Jeff knows that $k \geq 1$, since Travis knows that Jeff can see Nick. But Travis does not know that Nick knows that Jeff knows (or that Jeff knows that Nick knows) that $k \geq 1$ because Travis does not know his own eye color, and Travis knows that neither Nick nor Jeff knows his own eye color. Making the same arguments from Jeff's and Nick's perspectives, we now we have $k \geq 1$, $E(k \geq 1)$, and $E^2(k \geq 1)$, but we do *not* have $E^3(k \geq 1)$, which is what is really needed to instigate the suicides. The fact that no suicide occurs after the second day cannot be used by any blue-eyed islander to deduce his own eye color, so:

Theorem 3 If $k = 3$, no suicides will occur.

The logic gets even more convoluted as we progress to higher levels of complexity. Let it suffice to say that the logic continues. For any number of k blue-eyed islanders, observation alone is enough to achieve $E^{k-1}(k \geq 1)$, but *not* $E^k(k \geq 1)$. The latter is necessary to instigate the suicides, and it is provided under $C(k \geq 1)$, the common knowledge established by the announcement.

What do we learn from this? While the explorer's announcement doesn't offer any new information in its substance, what matters is its public nature. The public nature of the announcement allows each islander to deduce what kinds of propositions the other islanders are capable of making: Travis knows that Nick knows that Jeff knows—and so on.

Suppose the explorer's announcement were dispersed nonpublicly: The explorer secretly passed a slip of paper to each islander, letting each islander believe that he was the only one receiving the slip. Suppose the slip contained the same verbage as the otherwise public announcement (essentially, that $k \geq 1$). In this case, the islanders would continue living harmoniously, so long as $k > 1$.

Alternatively, suppose that just *one* of the blue-eyed islanders was known to be deaf and could not hear the public announcement. The society would indeed continue harmoniously, without suicide. (The public announcement would still cause suicides, though, if the deaf islander had brown eyes.)

Assuming that the public announcement is made and everyone hears it, do 100 blue-eyed islanders really need to wait 100 days to kill themselves, given that each can see 99 other blue-eyed islanders? In other words, each knows that 98 days will pass without suicide for sure. Can't they skip those 98 days? Either $k = 99$ or $k = 100$; there are only two possible states, so can't they deduce the necessity of suicide after a mere 2 days?

The 98-day lead time is indeed necessary to increase each islander's knowledge of each islander's knowledge of each islander's knowledge (and so on). Each day in the 98-day lead-up does not change the fact that each islander knows that $k = 99$ or $k = 100$. But each passing day does symmetrically confirm what each islander believes his fellow islanders can deduce (that his fellow islanders can deduce, and so on).

If the explorer were to realize his faux pas immediately after making his public announcement, he could still prevent a mass suicide by pointing to a blue-eyed islander and declaring: "*That* islander has blue eyes." The designated islander would have to commit suicide the next day, but none of the others would have to. The inaction of other blue-eyed islanders would no longer serve as an indicator of states of knowledge (about states of knowledge...), and thus no islander could deduce his own eye color. The tribe would return to a harmonious state of deep knowledge about its own ignorance.

8.3 Bottle Imp

The following paradox was introduced in an 1891 short story by Robert Louis Stevenson called "The Bottle Imp."

A special bottle contains a little imp that will grant the owner of the bottle his or her every desire. But there's a catch. To die as the owner of the bottle is to ensure eternal damnation. A person remains the owner of the bottle unless he or she can sell it. It cannot be given away, thrown away, destroyed, or otherwise disposed of. Furthermore, if ownership is to be legally transferred, the bottle must always be sold at a monetary loss—that is, at a lower price than its previous price. And it can be purchased only using some positive number of coins. Fractional pennies are not allowed. The rules about the bottle are considered common knowledge, or at least they must be conveyed to any potential buyer.

Surely, no rational person would ever pay 1 cent for the bottle. The bottle could then never be sold for less; the owner would face eternal damnation.

But then, one never would pay 2 cents for it, knowing that no rational person ever would buy it for 1 cent.

The backward inductive logic continues. A person would never buy the bottle without thinking that he or she could sell it. If the bottle could not be sold for 1 cent, it could not be sold for 2 cents to any rational buyer, or 3 cents to any rational buyer, or $10 (1,000 cents) to any rational buyer, or for $1 million for that matter. There is no price at which a rational person would buy the bottle, because there is no price at which it could be sold.

And yet, for a high enough price, it does not seem too unwise to purchase the bottle. Surely, if far enough removed from the lowest possible price of 1 cent, the bottle is worth purchasing. After all, the imp will grant the owner his or her every desire.

Claim 1
If a buyer is far enough removed from 1 cent, the bottle is a wise purchase.

Claim 2
The bottle is never a wise purchase.

Discussion and Resolution
The answer depends on how the economy is modeled. A variety of possible assumptions and structures are described here.

First, who are the players in this economy? Are we assuming a finite number of people? A countably infinite number of people? A continuum of people whose individuals can be represented by points on the real line?

Second, what are the payoffs? How bad is eternal damnation relative to the pleasure derived from owning the bottle? Is eternal damnation infinitely bad from a utility standpoint? Is the pleasure from owning the bottle infinitely good? Or is one finite and the other infinite? Can one own the bottle twice and derive additional pleasure from it, or does it have one-time ownership value?

Third, do the people in this economy all have the same utilities of (feelings about) damnation and ownership, or do they differ? In economic language, do they have *homogeneous* or *heterogeneous* preferences? If preferences are heterogeneous, what is their distribution? How can each type be described proportionally?

Fourth, what are the mechanics of a sale, and what is the matching process between a seller and potential buyers? Does the seller post a price and randomize over any interested buyers? Does the seller auction off the bottle? If so, what are the rules of the auction?

Fifth, how do we model time and the probability of death? Is time continuous or discrete? Do we assume that each person in the economy lives the same amount of time, or does death follow a stochastic process?

Sixth, is the behavior of all the people determined *endogenously*—that is, within the system according to optimal behavior? Or is there a segment of the population to which we can attribute an *exogenous* (fixed and external) demand curve? Such a distinction might help us allow some portion of the population to act irrationally.

Seventh, how much do the people in this economy know about the economy? Is everything common information?

These and other questions all matter. By choosing different assumptions, you can create an economic model to support either Claim 1 or Claim 2.

Claim 2 will tend to be supported by the following assumptions: (1) All players in the economy consider eternal damnation as infinitely bad, and owning the bottle as finitely good. (2) All people in the economy are rational, and they get to exercise free will. (3) All these assumptions are common knowledge. These are the conditions needed to support the backward inductive logic proving that no one will ever want to buy the bottle for any price.

8.4   The Raven

The Raven Paradox was put forward by the German-born philosopher Carl Hempel in the 1940s. It demonstrates a contrast between logic and intuition. Before we get to the paradox, it will be important to understand a rule of logical inference called *contraposition*.

Contraposition

In logic, one can express statements in the form "If *p* is true, then *q* is true." Statements in this form are called *conditional* statements and are notated this way:

$$p \rightarrow q$$

Another way of saying this is "*p* implies *q*." As an example, suppose *p* is the antecedent "It is raining," and *q* is the consequent "the ground is wet." Then $p \rightarrow q$ can be interpreted as "If it is raining, then the ground is wet," or equivalently, "A state of rain implies a wet ground."

The symbol for negation is ¬. Negation is the operation in logic that turns a statement into its opposite with a "not." If *p* expresses "It is raining," then ¬*p* expresses "It is not raining."

The *contrapositive* of a conditional statement is a new conditional statement obtained by negating both sides of the original and flipping them with each other. The contrapositive of $p \rightarrow q$ is

$$\neg q \rightarrow \neg p$$

This translates to "If the ground is not wet, then it is not raining." Conversion of a conditional statement to its contrapositive is called *contraposition*. It is a valid form of logical inference: A conditional statement is logically equivalent to its contrapositive.

It may be worth taking a few minutes to convince yourself of this. It will be easier to understand what follows if you are comfortable with contraposition. One way to ease yourself into the idea is to invent a few truthful conditional statements, form their contrapositives, and verify that the contrapositives are still true. Once you have verified that the exemplary conditionals are equivalent with their contrapositives, ask yourself why. Then you can generalize and reach the understanding that a conditional statement is *always* equivalent to its contrapositive.

You can also understand the concept visually. Suppose you shoot an arrow at a target that contains a bull's-eye. It would be true to say: "If you

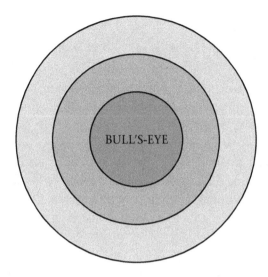

Figure 8.3
A target with a bull's-eye.

hit the bull's-eye, you have hit the target." It would also be true to say:
"If you have not hit the target, you have definitely not hit the bull's-eye."
In figure 8.3, try drawing a point that is within the bull's-eye but not on
the target at all. Impossible. You can't hit the bull's-eye without hitting the
target.

Any conditional statement $p \to q$ can be visualized in a similar way: as a
Venn diagram, with p being a circle contained within q (or equal to it). For
example, consider: "If an animal is a dolphin, then it is a mammal." Here,
q is the big circle of all mammals, and p is a circle inside q representing
dolphins. It is impossible to point to an entity on this diagram that is not a
mammal and is a dolphin.

You can never be outside the bigger circle *and* inside the smaller circle.
That is the rule of contraposition.

The Paradox

Now, here comes the paradox. Consider the statement "All ravens are black."
This can be rephrased as a conditional statement:

> If a thing is a raven, that thing is black.

If this statement is a hypothesis but not a certainty, the hypothesis is supported by any sighting of a black raven; a sighted black raven is evidence.

The contrapositive of this statement is

> If a thing is not black, that thing is not a raven.

If this statement is a hypothesis but not a certainty, the hypothesis is supported by any sighting of a nonblack nonraven; a sighted nonblack nonraven is evidence. A red strawberry, for example, provides evidence in support of the contrapositive.

The original conditional statement and its contrapositive are logically equivalent. This means that a red strawberry gives us information about ravens, and in particular, evidence supporting the hypothesis that all ravens are black! How can this be, given that a red strawberry is neither a raven nor black?

Claim 1
A red strawberry does not provide evidence in support of the hypothesis that all ravens are black.

Claim 2
A red strawberry does provide evidence in support of the hypothesis that all ravens are black.

Discussion and Resolution
Recall (from chapter 4 on probability) Ayn Rand's definition of a concept as a "mental integration of two or more units which are isolated by a process of abstraction and united by a specific definition."

Before anything can be said or asked about ravens, there must be a well-formed concept of "raven," and a clear understanding of the scope of its definition. If blackness is integral to the concept of "raven," then the definition itself tells us that all ravens are black, and there is no uncertain hypothesis to support or undermine with evidence. The only relevant pieces of "evidence" are those that serve as the basis for concept formation: the individual birds with some similar properties, which are perceived and integrated into a useful concept. It is by this process of concept formation that humans can organize and understand the world they live in.

A question that has challenged philosophers for millennia is how to handle *edge cases*—individual units for which there are arguments to admit or reject membership under the concept. One such thought experiment is called Theseus's Paradox. Suppose that a ship sailed by Theseus is kept in a museum. Over time, rotting parts are replaced in such a way as to preserve the ship's original appearance. After some time, all the original parts have been replaced. Is the restored ship still Theseus's ship?

Here, the concept "Theseus's ship" is made ambiguous by this backstory. However, the ambiguity can be handled through further specification: There is "Theseus's original ship," and there is "Theseus's restored ship," each a new, unambiguous concept that employs a new property in its definition (original versus restored).

Suppose you define a raven as follows:

> A *raven* is a bird that resembles a black crow, is slightly larger than a crow, and has a harsh call.

Walking through the forest one day, you encounter a bird, which you name Merlin. Merlin resembles a crow in shape, is slightly larger than a crow, and has a harsh call. However, Merlin is white. Is Merlin a raven? You have three alternatives:

- Decide that Merlin *is* or is *not* a raven, because Merlin is sufficiently characteristic or uncharacteristic of ravens according to the existing definition.
- Disambiguate the concept of "raven" by updating your definition either as "a bird that resembles a crow in shape, is slightly larger than a crow, and has a harsh call," or as "a black bird that resembles a crow in shape, is slightly larger than a crow, and has a harsh call." Then include or exclude Merlin accordingly.
- Disambiguate by specifying a new concept like "white raven."

The paradox at hand—the paradox of confirmation—applies only when the definition of the "raven" concept does not encompass coloration. If "All ravens are black" is a hypothesis of which we are uncertain, then we have already presupposed a definition of "raven" that is understood to be independent of blackness. That is, we have presupposed that a bird can be black or nonblack and still be considered a raven. Otherwise, we would have certainty, not a hypothesis.

Generally, the property being tested must not be integral to the definition and the definition must be well specified outside this property. Only then does it make sense to hypothesize whether some p implies some q. If p and q are known to be related by definition, we already have our answer.

Moving on, we therefore assume a definition of "raven" that can admit both black and nonblack birds. Assume also that more than one raven and more than one nonblack object are known to exist; otherwise, if we know there is only one raven or only one nonblack object, the hypothesis is easily tested.

Next, we need to know what constitutes evidence in support of the hypothesis that that all ravens are black. Consider two distinct definitions of *evidence* in support of the hypothesis:

- *Definition 1: Evidence* is any object that is consistent with (i.e., does not contradict) the hypothesis that all ravens are black.
- *Definition 2: Evidence* is any object that is consistent with the hypothesis *and* whose sighting increases the probability that all ravens are black.

Let's first look at Definition 1 as a concept of "evidence." Indeed, any black raven, and any nonblack nonraven, supports the hypothesis as evidence. However, the hypothesis can be proved true only if *all* ravens are observed to be black, and the equivalent contrapositive can be proved true only if *all* nonblack things are observed as nonravens. Any single piece of evidence in a category is necessary but insufficient to prove the hypothesis. If it feels counterintuitive that a red strawberry should qualify as evidence, some comfort may be taken in the fact that while that strawberry's nonravenhood is necessary, it is insufficient to prove the hypothesis by contraposition. This is an easier pill to swallow.

Now let's try Definition 2 of the "evidence" concept, which is stricter and requires an increase in probability of a true hypothesis. A number of Bayesian-style solutions have been proposed, which show how observations of black ravens and nonblack nonravens affect the probability of the hypothesis. Generally, Bayesian solutions require some assumptions with respect to background information. For example, we may need to assume some finite number of ravens, and some finite number of nonblack things. We may also need some assumptions about the probability that any given thing is black, or is a raven.

Examples of Bayesian models are left out of this book; the details of their setup are less important at a high level. The intuition they generally confirm is that observing a black raven supports the hypothesis much more strongly than observing a nonblack nonraven. The observation of a nonblack nonraven has an exceedingly small, though positive (again, depending on assumptions) impact on the probability that the hypothesis is true. And once again, in most Bayesian solutions, a single observation is necessary but insufficient to prove the hypothesis. While it feels like the hypothesis can be supported gradually, either all ravens or all nonblack things must be observed for absolute proof.

The British mathematician Jack Good demonstrated the importance of background information in setting up a model. He proposed a context in which the observation of a black raven actually could reduce the probability that the hypothesis is true:

> Suppose that we know we are in one or other of two worlds, and the hypothesis, H, under consideration is that all the crows in our world are black. We know in advance that in one world there are a hundred black crows, no crows that are not black, and a million other birds; and that in the other world there are a thousand black crows, one white one, and a million other birds. A bird is selected equiprobably at random from all the birds in our world. It turns out to be a black crow. This is strong evidence … that we are in the second world, wherein not all crows are black. Thus the observation of a black crow, in the circumstances described, undermines the hypothesis that all the crows in our world are black. Thus the initial premise of the paradox of confirmation is false, and no reference to the contrapositive is required.

Having considered all these factors, let's revisit our paradox of confirmation, the notion that observing a red strawberry provides information about ravens. Is this notion true? You can see now that the answer relies on a number of factors. To begin, we must have a clear concept of "raven" that is independent of color. We also must have a clear concept of "evidence," of which there are multiple reasonable options. If the concept of evidence involves probability, then we must examine the implications of any choice in background information and prior knowledge. Other factors can affect the validity of the notion, too. As factors like these are introduced and clarified, the paradox of confirmation—if one exists—becomes less of a paradox. It begins to jibe with intuition.

By a definition of "raven" that does not include color, and by a definition of "evidence" that requires necessity but not necessarily sufficiency, yes, it is possible for a red strawberry to support the hypothesis that all ravens are black. But other situations can be constructed that imply the opposite.

9 Geometry

You can't criticize geometry. It's never wrong.

—Paul Rand

Geometry is the study of points, lines, surfaces, solids, and generalizations of these in higher dimensions. Its original scholar, considered the "father of geometry," was the ancient Greek mathematician Euclid of Alexandria, who lived circa 300 B.C. Euclid championed the idea of the rigorous proof. His magnum opus was the book *Elements*, which is still used to teach mathematics today. Euclid formally defined many geometric and other mathematical concepts more than two millennia before Georg Cantor established the modern field of set theory in the late 1870s. Set theory provides a language that can be used to define and describe most mathematical objects. Euclid's accomplishments are particularly impressive given that he did not have this modern language.

In fact, some of Euclid's definitions hint at the notion of sets and point to the importance of considering collections of objects. In Book I of *Elements*, Euclid defines a circle as a shape living in a plane with a certain point, called the *center*, and a boundary, such that all straight lines from the center to the boundary have equal length. That's not much different from the way we think of circles now, as the set of all points in a plane that are equidistant from the center (or of a disk, which includes all the points from the center to the boundary).

The ancient Greeks did not have calculus, either, nor did they have the understanding of real numbers that we do. Yet Euclid still was able to prove, using limits, that the area of a circle is proportional to the square of its radius. Other ancient Greeks, including Archimedes, achieved similar proofs.

Now that we *do* have tools such as calculus, set theory, and real analysis, our understanding of geometric objects has deepened. Yet there are still ways to find psychological discord in the realm of geometry.

9.1   Fractional Dimensions

A point has zero dimensions (0D). A line has one dimension (1D). A square has two dimensions (2D). A cube has three dimensions (3D). Higher-dimensional objects can be described mathematically. What do these dimensions all have in common? They are *integers*.

What if a stranger told you that certain theoretical objects can be conceived that have fractional dimensions, like 1.5 dimensions? Can such objects exist in theory, or are they reserved for fairy tales? Their existence would be paradoxical, indeed.

Claim 1

Dimensions can have only integer values. We know this because we can describe the location of any point on an object using coordinates. The object's dimension is the minimum number of parameters or axes (like up-down, left-right, forward-backward) needed to express all such coordinates. You can't have a fractional parameter or axis. So you can't have a fractional dimension.

Claim 2

Dimension is another word for the concept "aspect," "feature," or "extent." There is a useful concept of dimension that permits fractions.

Discussion and Resolution

Claim 2 is correct.

The intuitive concept of dimension is the one presented in Claim 1. We tend to think of an object's dimension as the number of parameters required to locate any unique point of that object. But as discussed in section 1.3 in chapter 1, any point specified by *n* parameters can also be specified by just 1 parameter if the digits of the coordinates are interwoven. This is true because continuous lines, planes, solids, and higher-dimensional objects possess the same cardinality. "The number of parameters required to specify any point," therefore, is not a good definition of dimension.

A better concept is the *Lebesgue covering dimension*, which provides a rather complex definition that aligns with the intuitive concept of dimension, in that it assigns 1D to lines, 2D to planes, 3D to solids, and so on. The formal definition of the Lebesgue covering dimension is complicated, so it is omitted from this book. It involves covering objects with other sets and then refining the way those objects are covered. This concept is one of multiple definitions of dimension in *topology*, the branch of mathematics that studies the properties of objects that are preserved under bending, twisting, crumpling, and stretching, but not by tearing and gluing.

A sphere is a 3D object whose boundary is a 2D surface. A disk is a 2D object whose boundary is a 1D line (a circle). A line segment is a 1D object whose boundary consists of two 0D points. We can, and sometimes must, think of objects as being embedded in spaces of higher dimension than their own. Visualize the surface of a sphere, for example. You are imagining the surface, a 2D object, in 3D space. In fact, you *have* to. You need 3D to imagine the sphere's surface in its entirety. So how can the surface be a 2D object?

What matters is not the global fact that the sphere's surface exists in 3D space, but rather the fact that the surface looks 2D at each of its points locally. Imagine an ant walking on the surface of a beach ball, for example. No matter where the ant is on the surface, locally, it has only two axes (left-right, forward-backward) of movement. Nonetheless, an observer watching the ant crawl along the beach ball sees the ant traversing three axes of movement (x, y, and z). The ant may be fixed to a 2D object, but it is moving in 3D space. (An object that looks n-dimensional in the immediate vicinity of any of its points is called a *pure n-manifold*. The surface of a sphere qualifies as a pure 2-manifold.)

Let's introduce an informal concept of relative *mass* to our mathematical objects. Suppose that geometric lines are made of infinitely thin 1D wire, and that surfaces and solids are made of 2D sheets and 3D chunks of metal. Furthermore, let these materials have some mass. Of course, nothing in real life is infinitely thin or infinitely flat, so it's strange to talk about the mass of an infinitely thin wire or infinitely flat plane. But this is a mathematical manufacturer, not a physical manufacturer, so some abstract notion of mass can apply. Let the materials (wire, sheets, chunks) have uniform density. That means that twice some length, area, or volume has twice the mass. We will only need to use this concept of mass in relative terms, comparing masses of things; we will never use it in absolute terms.

Let's also suppose we can scale an object. *Scaling* an object means making it bigger by stretching it in some direction while maintaining its proportions.

What happens to the mass of a 1D object as we scale it? Its mass increases proportionally with the scale (to the power of 1).

What about 2D objects, like squares? If we scale a square by a factor of 2, doubling its side length, we end up with something that looks like four of our original squares stuck together. The mass has increased by four times: the scale factor, to the power of 2.

What about 3D objects? If we scale a cube by a factor of 2, doubling its side length, we end up with something that looks like eight of our original cubes stuck together. The mass has increased by eight times: the scale factor, to the power of 3.

Notice a pattern? Scaling a line, square, or cube creates a new object whose mass increases by the scale factor raised to the power of the dimension:

$$\text{Mass factor} = (\text{Scale factor})^{\text{Dimension}}$$

Isolating dimension, we obtain

$$\text{Dimension} = \frac{\log(\text{Mass factor})}{\log(\text{Scale factor})} \qquad (9.1)$$

This relationship is an informal definition of the *Hausdorff dimension* for self-similar objects, named after Felix Hausdorff. For any traditional geometric object, the Hausdorff dimension will agree with the topological dimension. However, the concepts mismatch in the context of more complex objects. Let's take a look at a few self-similar patterns that replicate themselves when examined at any scale. Such patterns would look the same from a bird's-eye view as they would under a microscope. We will build these objects using iteration.

Cantor Set

The first mathematical object we'll examine is the Cantor set. This object has a number of remarkable properties and is named after Georg Cantor, although the set was described by other mathematicians before him.

It is constructed as follows. Start with a line segment of some fixed length. You are now at Level 0 of our iterative algorithm.

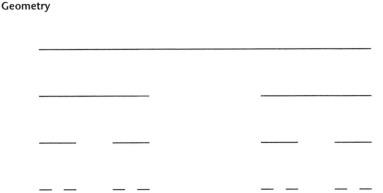

Figure 9.1
Levels 0–3 of the Cantor set.

Now remove the middle third of the line segment, but keep all end points. You are left with two disjoint line segments. You are now at Level 1 of our iterative algorithm.

Repeat the process by removing the middle third of each of the remaining (two) line segments, without removing any end points. You are now left with four disjoint line segments. You are now at Level 2 of our algorithm.

Repeat the process by removing the middle third of each of the remaining (four) line segments, and so on. The iterations continue ad infinitum; Level n approaches infinity.

This iterative process constructs the Cantor set (see figure 9.1).

The measure of length (called a *Lebesgue measure*) of the original set was positive, and with each iteration, it diminished until reaching an eventual measure of zero. In other words, the sum of the lengths of the remaining line segments approaches zero, and in the limit, there is no continuous line segment remaining with positive length.

And yet it can be proved that the Cantor set is uncountable. The cardinality of the final set of points is the same as the cardinality of the set of points constituting the continuous line at Level 0. The measure has gone to zero, but the number of points has stayed the same! There are no lines remaining, but there are an uncountable number of remaining "end points."

The Cartesian product of the Cantor set with itself (a *Cartesian power* of the Cantor set) generalizes the Cantor set into higher dimensions, leaving us with so-called *Cantor dust*.

Notice that in the process of creating the Cantor set, we actually create an infinite number of Cantor sets. The line at Level 0 gets broken into two lines at Level 1, each of which becomes its own Cantor set spanning one-third the length of its parent line. The process continues. If we were to take any Cantor set and scale it up by a factor of 3, we would end up with two Cantor sets similar to the one being scaled—twice the original mass.

The Hausdorff dimension of the Cantor set, therefore, is

$$\text{Dimension(Cantor set)} = \frac{\log(2)}{\log(3)} \approx 0.63$$

A fraction!

Sierpiński Carpet

Described by the Polish mathematician Waclaw Sierpiński in 1916, the Sierpiński carpet is a 2D analog of the Cantor set. Though not a Cartesian power of the Cantor set (and hence not Cantor dust), it's an easier version to visualize and has the same properties of interest. (Carpets and dust go hand-in-hand in mathematics, just as in real life.)

To construct the Sierpiński carpet, start with a solid, filled-in square (Level 0). Now divide the square into a 3×3 grid of nine smaller, equal-sized squares and remove the middle square, but not its boundary points (Level 1). Do the same for each of the remaining smaller squares (Level 2). Then do it again, and again, ad infinitum, with Level n approaching infinity. The process is shown in figure 9.2.

After performing this iteration an infinite number of times, you are left with the Sierpiński carpet: an object with zero area but an infinite boundary length (including inner perimeters).

After each iteration, the remaining area is equal to $\frac{8}{9}$ of the previous area. The area shrinks in constant proportion, and so its limit as n approaches infinity is zero. The perimeter, however, increases without bound. Figure 9.3 depicts the Sierpiński carpet after eight iterations.

The Sierpiński carpet is actually an arrangement of eight smaller Sierpiński carpets (implying a mass factor of 8), each of which is one-third the size (scale factor of 3). The Hausdorff dimension, therefore, is

$$\text{Dimension(Sierpiński carpet)} = \frac{\log(8)}{\log(3)} \approx 1.89$$

Notice how the higher-level image starts to look gray. This is itself a kind of paradox—every point in the square should be pure black or pure white. There are two reasons for this. One has to do with the finite nature of digital images, and the other has to do with the limitations of the human eye.

Figure 9.2
Levels 0–3 of the Sierpiński carpet.

Assume a computer program generates the image by running through a desired n iterations. As we increase n, making the image truer and truer, at some point the mathematical demands will exceed the possible pixel granularity of the image. Inevitably, after enough iterations, there will be individual pixels, which can have the property of only one uniform color, which need to be divided into subsets of both pure black and pure white. The computer, which can assign only one color to a pixel, needs some sort of instruction for handling this problem. One option is to force the computer to stick to pure black or white, but then the picture loses its mathematical integrity very quickly. Another is to choose some shade of gray that reflects the appropriate proportion of black and white as a function of n. For example, if some pixel were supposed to become one part pure white and eight parts pure black, then the whole pixel becomes a single shade of dark gray, which could be described as one part white and eight parts black. This is a

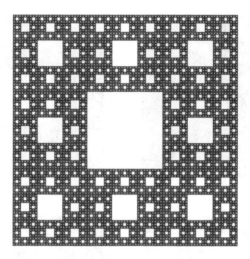

Figure 9.3
Level 8 of the Sierpiński carpet.

common way to resize an image. Reducing the number of pixels means that certain pixels take the average of the colors around them.

There's a second reason the image looks gray, though: Your eyes aren't perfect. Let's say you actually could print the image with perfect fidelity, and each pixel truly were a point. Suppose you watch the image being generated iteration by iteration. With each iteration, it becomes a better approximation of the true set. In the first iteration, the strokes of black ink are large and thick. But with each successive iteration, the strokes of black ink become thinner and thinner. Eventually, the thickest one is hair-thin, then atom-thin, and so on. The image becomes more and more dominated by white space. At some point, the human eye can no longer perceive individual strokes. It can only perceive averages, and that average tends ever more toward pure white. So as n approaches infinity, the average color perceived by your eye approaches pure white. After an infinite number of iterations, the black area would have measure zero, no ink would be necessary, and long before that time, your eye would see nothing on the page. For the same reason, you don't see the bacteria on the surfaces around you, even though bacteria come in all kinds of bright colors.

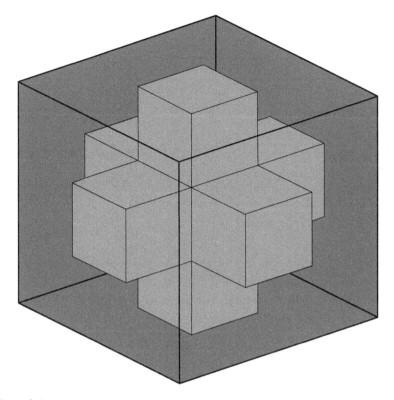

Figure 9.4
Level 1 of the Menger sponge.

Menger Sponge

Described by the Austrian-American mathematician Karl Menger in 1926, the Menger sponge is a 3D analog of the Sierpiński carpet (not to be confused with the similar, but nonetheless different, Sierpiński sponge). To construct the Menger sponge, begin Level 0 with a cube. The first level is achieved by dividing the cube into $3 \times 3 \times 3$ (a pattern of smaller cubes that looks like a Rubik's Cube) and drilling holes through the middle square of each face, as shown in figure 9.4.

Each subsequent iteration performs the same action on the remaining pieces: Divide each cube into twenty-seven smaller cubes and drill through the central square along each axis. Every face of the cube becomes a Sierpiński carpet. Somewhat paradoxically, the volume of this object, through its iterative construction, approaches zero, and its surface area

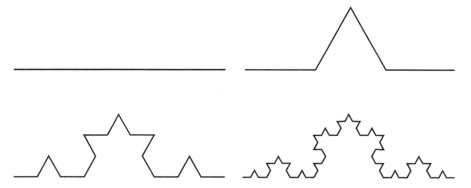

Figure 9.5
Levels 0–3 of the Koch snowflake.

approaches infinity. Yet upon completion, it is neither a solid nor a surface, but a "curve" having topological dimension of only 1.

Each Menger sponge contains twenty cubes that are one-third its scale. The Hausdorff dimension, therefore, is

$$\text{Dimension(Menger sponge)} = \frac{\log(20)}{\log(3)} \approx 2.73$$

Koch Snowflake

The Koch snowflake, described by the Swedish mathematician Helge von Koch, is constructed as shown in figure 9.5. Begin with a line segment. Replace its middle third with two line segments, joined at the top and respectively connected at the bottom, each equal in length to one-third of the original segment. Repeat on each remaining segment ad infinitum. The snowflake appearance is created by arranging three copies of the result in a triangle.

The result is a curve of infinite length. Not only that, but the line connecting any two points along the Koch snowflake has infinite length.

Each Koch snowflake contains four snowflakes that are one-third its scale. The Hausdorff dimension, therefore, is

$$\text{Dimension(Koch snowflake)} = \frac{\log(4)}{\log(3)} \approx 1.26$$

Fractals in General

All the patterns described here belong to a class of objects known as fractals. The mathematician Benois Mandelbrot coined the term in 1975 from the

Latin root *fractus*, meaning fractured or broken up. He defined a *fractal* as an object whose Hausdorff dimension exceeds its topological dimension. He subsequently expanded the definition to include any shape made of parts that are somehow similar to the whole.

Fractals don't have to be exactly self-similar, like the ones described in this section. Neither is the Hausdorff dimension their only concept of dimension. There are other concepts as well, including ones that can assign dimension to rough patterns observed in nature, like coastlines, mountains, forests, and wavy ocean surfaces, using statistical methods.

The Hausdorff dimension is really a measure of roughness and jaggedness, a concept that allows rough objects to straddle dimensions and take noninteger values. Some infinitely iterated processes can even take an object from one dimension to the next. Space-filling curves, such as the Peano curve or the Hilbert curve, start with a line segment that is iteratively lengthened and bent until it occupies the area of a square.

Visually, fractals can be quite stunning. Mandelbrot summed them up concisely: "Beautiful, damn hard, increasingly useful. That's fractals."

9.2 Aristotle's Wheels

The Aristotle's Wheels Paradox appeared in the Greek text *Mechanica*, whose authorship is disputed but generally attributed to Aristotle. Imagine two concentric wheels: one inside the other, with a shared center point. Now roll the outer wheel in a line, for a distance equal to the outer wheel's circumference (see figure 9.6). Every point on the perimeter of the outer wheel will touch the ground exactly once; there is no slipping. The inner wheel traces out a path of equal length, and every point on its perimeter also touches the ground exactly once. But this implies that the inner wheel,

Figure 9.6
Aristotle's wheels.

which is smaller, must have the same circumference as the outer wheel—a contradiction.

Claim 1
It must be that the inner wheel has the same circumference as the outer wheel.

Claim 2
The lines traced by the wheels have equal length, and each point on the perimeter of the wheels touches the ground exactly once, but these facts do not imply that the wheels have the same circumference.

Discussion and Resolution
Claim 2 is correct. The two wheels rotate with the same angular velocity (number of degrees per second) and travel with the same horizontal velocity. What differ are the ratios of the circumferences to the horizontal distance traveled. If gears of equal size were placed along the horizontal lines and around the wheels, then the gears would jam, given the difference in these ratios.

Two lengths can be different and still have the same cardinality—the same number of points in the sets defined by their lengths. If two sets have the same cardinality, then a *bijection*, or a one-to-one correspondence, can be drawn between them.

The horizontal distance and the circumference of the outer wheel have the same length, so a bijection between them is easy to imagine. The horizontal distance and the circumference of the inner wheel do not have the same length; they can be bijected in the manner illustrated in figure 1.1 in chapter 1. This is the bijection made visible by the action of rotating the wheels along the horizontal path.

9.3   Coin Rotation Paradox

Imagine placing two of the same coins beside each other on a table. Now rotate one of the coins (Coin B) around a stationary coin (Coin A) without allowing any slipping of the coins (figure 9.7). Notice that the rotating coin completes *two* full rotations about its center, not just one. At the same time, each point along the circumference of Coin B touches

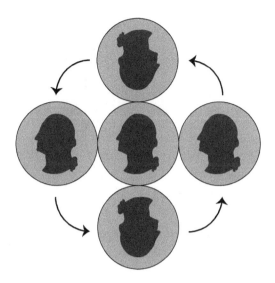

Figure 9.7
One coin rotating around the other.

the perimeter of Coin A exactly once, not twice. This seems to imply a contradiction.

Claim 1
It must be that the coins are somehow slipping against each other, or that the points on the perimeter of Coin B are touching Coin A twice, not once.

Claim 2
Coin B's rotation can be broken into two separate motions, which explain the 720, not 360, degrees of rotation.

Discussion and Resolution
Claim 2 is correct. Coin B has two separate axes of rotation. First, imagine what it would look like to rotate Coin B around Coin A, while keeping George Washington's face on Coin B looking at the center of Coin A at all times. In other words, the same point on the perimeter of Coin B remains tangent to Coin A as it slides in rotation. After returning to its original spot, Coin B would have completed one rotation. To prevent sliding, we add another axis of rotation to Coin B, about its own center. This makes two axes of rotation.

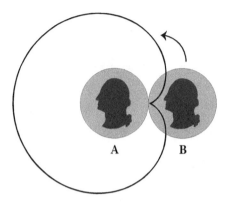

Figure 9.8
The cardioid path drawn by the initial tangent point of Coin B.

If you were to follow any point on the perimeter of Coin B through the entire process, you would find that each point does indeed make contact with Coin A only once. The initial point traces what is known as a *cardioid* path, a shape reminiscent of a heart (see figure 9.8).

You can also imagine rotating Coin A and Coin B simultaneously, each about its own center, one clockwise and the other counterclockwise. Each coin would rotate 360 degrees before the coins would return to their original positions. The total degrees rotated would be 720. Two total rotations would have been completed between the two coins. Each point on any given coin would have touched the perimeter of the other coin exactly once.

Now, suppose that while doing this, you were to rotate the table upon which the coins are resting. Rotate the table about the center point of Coin A, with the opposite angular velocity to that of Coin A, so that Coin A appears stationary relative to the walls of the room. Coin B would now appear to rotate around Coin A as described in the original problem, covering the two full rotations itself. As before, no point on its perimeter would touch Coin A twice.

9.4 Roly-Poly's Staircase

A point-sized roly-poly is trying to climb up a straight, frictionless hill but keeps rolling back down. His friend the grasshopper offers to build a staircase up the hill using triangular wedges that magically stick to the hill's friction-less surface, giving the roly-poly a way to climb. But the roly-poly

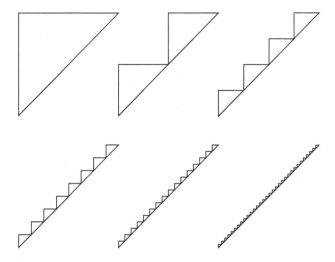

Figure 9.9
The roly-poly's staircase, created iteratively.

complains that the wedges add distance because each one carries him up and across, whereas the slope itself follows a straight diagonal. "Everyone knows the shortest distance between two points is a straight line," says the roly-poly. "No worries," says the grasshopper, "I'll just make my wedges smaller and more numerous, decreasing your average distance from the side of the hill." As the grasshopper builds his staircase with smaller and smaller steps, the roly-poly's average distance from the hill indeed approaches zero. But the roly-poly never seems to save any time. For some reason, the path up the stairs is always longer than the path straight up the hill. See figure 9.9 for an illustration of this scenario.

Claim 1

The stairway path converges to the hill path. If the grasshopper were to halve the perimeter of the wedges ad infinitum, the stairway would eventually smoothly coat the side of the hill, and the roly-poly would indeed save time.

Claim 2

As the stairs get smaller, the distance up the stairs stays constant; it does not converge to zero. At the same time, the average distance between the roly-poly and the nearest point on the hill converges to zero. This appears to be a contradiction, but it is not. These are two different series.

Discussion and Resolution

Claim 1 sounds tempting but is actually nonsensical. Consider the sentence: "The stairway path converges to the hill path." What does it actually mean for one path to converge to another? It does not make sense for one object to converge to another object; a stairway cannot be said to converge to a flat line. What we *can* do is look at two distinct series, as in Claim 2.

One series represents the total length of the staircase. We can define this total length as the sum of all vertical and horizontal distances comprising the staircase. Even as the grasshopper doubles the number of steps and halves the length of each, the sum of the individual lengths will remain constant. The total length of the staircase, as a function of the number of steps, does not converge to zero. The output of the function is constant and positive.

An entirely separate series to consider is the one generated by taking the average distance between the roly-poly and the nearest point on the hill. We can define this average distance as a function of the number of steps. As the number of steps approaches infinity, this function converges to zero. For any finite number of steps, however, the roly-poly's climb never follows a smooth path up the hill, nor does its average distance to the hill ever equal zero; it is always positive.

The key here is to distinguish these two series conceptually. They are not talking about the same thing. There is no reason to expect one to converge to the same value as the other.

9.5   Block Stacking

Suppose you have a set of n rectangular blocks, all equal in size, shape, and mass. Your objective is to stack them in such a way that maximizes the overhang of the top block; that is, you want to maximize the horizontal distance between the bottom block and the top one (see figure 9.10). As n approaches infinity, what is the maximum overhang that can be achieved?

The resulting stack must be stable. We will assume the simplest version of this problem, which does not allow counterbalancing. Blocks must be stacked simply one on top of another.

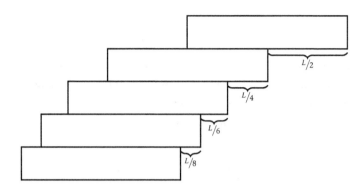

Figure 9.10
Stacking blocks.

Claim 1

As the number of blocks approaches infinity, the maximum overhang must approach some finite length. Otherwise, the stack of blocks would clearly topple.

Claim 2

The stack of blocks can in fact reach an arbitrarily large overhang. As the number of blocks approaches infinity, the maximum possible overhang approaches infinity as well.

Discussion and Resolution

Claim 2 is correct. If you had an infinite number of blocks, you could stack them in a way that achieves an infinite horizontal distance.

Let L be the length of each block. Start by considering the case with only two blocks. The top block's center of mass is at the point $\frac{L}{2}$, and so an outcropping of $\frac{L}{2}$ can be achieved.

What if you had three blocks? We know from the case with two blocks that the top two blocks can be stacked at a horizontal distance of $\frac{L}{2}$ from each other. The total horizontal length of those two blocks is $\frac{3L}{2}$, and the center of mass is a distance $\frac{3L}{4}$ inward from either side. The top two blocks, therefore, can hang over by $\frac{3L}{4}$ from the bottom block.

The same logic continues. In general, the maximum overhang that can be achieved with n blocks is given by

$$\text{Overhang}(n) = \frac{1}{2} \sum_{i=1}^{n} \frac{1}{i}$$

This value is equal to half of the nth partial sum of the so-called *harmonic series*. The nth term in the harmonic series is called the nth *harmonic number*. It can be shown that the harmonic series diverges—it approaches infinity. We can prove this by comparing the harmonic series to a known diverging series:

$$1 + \frac{1}{2} + \frac{1}{4} + \frac{1}{4} + \frac{1}{8} + \frac{1}{8} + \frac{1}{8} + \frac{1}{8}\ldots < 1 + \frac{1}{2} + \frac{1}{3} + \frac{1}{4} + \frac{1}{5} + \frac{1}{6} + \frac{1}{7} + \frac{1}{8}\ldots \quad (9.2)$$

The series on the left side of equation 9.2 here can be grouped in such a way that it obviously diverges: It is a sum of 1, plus an infinite number of 1/2s. Hence, it diverges to infinity. The series on the right, whose every nth term is equal to or greater than the nth term of the lesser series, therefore must diverge to infinity as well.

9.6 The Ant's Elastic Adventure

An ant begins a journey at a point on a rubber band whose circumference is 1 kilometer (see figure 9.11). The ant moves at a constant rate of 1 centimeter per second relative to the rubber it crawls on. At the same time that the ant begins its journey, the rubber band's circumference begins expanding at a constant rate of 1 kilometer per second. Will the ant ever make it all the way around the rubber band and reach the starting point again?

Claim 1
Of course not. The ant moves at 1 centimeter per second, but the path ahead of him grows at a much faster rate. The ant will never make it back to where he started.

Claim 2
It will take him a long, *long* time, but the ant will indeed come back to where he started.

Discussion and Resolution
Claim 2 is correct. It seems as though the ant should never be able to traverse the entire rubber band because the band's circumference increases at a rate faster than the ant's crawling speed. But when we express the ant's position

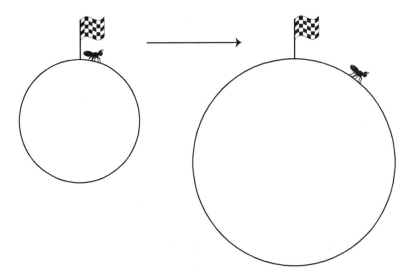

Figure 9.11
The ant on his adventure around the expanding rubber band.

on the band as a function of time, we find that our little bug will reach his starting point again someday.

Moving forward, let's imagine the ant moving along a flat, linear rubber rope rather than around a circle. The mathematics are the same but easier to describe linearly. The question now becomes whether the ant will ever reach the end point of the ever-stretching rope.

Suppose the rope's initial length is l_0, and the rope stretches at a constant rate of v meters per second. At time t, the rope's length is $l(t) = l_0 + vt$. Points along the rope move proportionally with the end point. The starting point stays fixed, and every point a fraction of the way toward the end point moves at the same fraction of the end point's velocity. A point on the rope that is a distance X from the starting point moves at a velocity $\frac{vX}{l_0+vt}$.

Suppose the ant's constant velocity relative to the rubber he crawls on is a, in meters per second. Let the position of the ant, relative to the starting point, be given by $y(t)$. The ant's velocity relative to the starting point is the derivative of this—the change in his position with respect to time, or $y'(t)$. At any given time, the ant's velocity relative to the starting point is the sum of his own velocity (relative to the rubber) plus the speed at which the expanding rope carries him. We can express this relationship analytically:

$$y'(t) = a + \frac{vy(t)}{l_0 + vt} \tag{9.3}$$

This is a differential equation, an equation that relates a function with its own derivative. The step-by-step solution to the equation is omitted here, but it yields the following equation for the *time* it takes the ant to reach the end point:

$$T = \frac{l_0}{v}\left(e^{\frac{v}{a}} - 1\right)$$

This value is finite for all $l_0, v > 0$, and $a > 0$. Plugging in the initial values, we obtain $T = (e^{100,000} - 1)$ seconds, or approximately $2.81 \times 10^{43,429}$ seconds, which is $6.85 \times 10^{43,411}$ times the age of the universe.

Assuming that an ant eats a third of its body weight each day and weighs approximately 3 milligrams, this brave ant of ours would need to eat approximately $8.89 \times 10^{43,415}$ kilograms to power its journey.

Alas, the observable universe is estimated to contain ordinary matter whose total mass equals 10^{53} kilograms. All the dust, gas, moons, planets, stars, and black holes in the approximately 100 billion galaxies would make but the smallest appetizer relative to the ant's full dietary needs on his journey—and this would remain true even if the ant could tap into nuclear, not merely chemical, energy.

But mathematically speaking, he'd get there eventually.

10 Operations

Numbers have life; they're not just symbols on paper.
—Shakuntala Devi

The extent to which we understand concepts is often the extent to which
we use them. After all, the concepts we form serve a practical purpose:
They help us grasp and organize the world around us. We form concepts
by observing certain similarities between things, or units, and uniting
those things under a specific definition. Concepts are the building blocks
of knowledge, and for human beings, they are a requirement for survival.
We must integrate the sensory inputs provided to our brains into concepts
before we can identify the values that lead to our flourishing.

We usually form concepts in ways that help us meet the needs of daily
life. Sometimes, however, an unusual context will arise such that an existing
concept—which seems like it ought to apply—doesn't quite manage to do
so. In such cases, we have the choice to refine or reformulate the existing
concept or create a new one.

This chapter examines some of the operations you use regularly in math:
arithmetic operations and differentiation. You probably have useful, working concepts of these operations. When faced with the task of performing
them, you know what to do. They seem easy enough. But how deeply do
you understand these operations? Do you perform them mechanically, or
do you have a fully integrated concept of what you are doing?

The following paradoxes are intended to challenge your concepts of some
basic mathematical operations—in particular, what it means to add, divide,
and take derivatives. The final two sections will take you through a process

of extending and generalizing the concept of summation, enabling you to "add" things that seemed previously unsummable.

10.1 Missing Dollar Riddle

Three guests check into a hotel, where the receptionist tells them the cost is a total of $30, or $10 per person. After the guests pay, the receptionist realizes she has made a mistake. The total bill was actually meant to be only $25. The receptionist gives the bellhop a $5 bill and instructs the bellhop to reimburse the difference to the guests. Realizing that he cannot divide $5 equally, the bellhop decides to give $1 to each of the guests and pocket the remaining $2 himself as a tip.

In the end, each guest paid $9—the initial $10, with a reimbursement of $1 each. The three guests, therefore, paid a total of $27 for the rooms. Add in the bellhop's $2 tip, and the total becomes $29. But the initial amount handed over by the guests was $30, so where did the remaining $1 go?

Claim 1
The receptionist must have somehow stolen $1.

Claim 2
There is no reason why the revised room cost, plus the tip, should add up to $30.

Discussion and Resolution
Claim 2 is correct. If we properly account for every dollar of the original $30, we find no missing money. Of the original $30, there is now $25 in the cash register, $1 in each of the three guests' pockets, and $2 in the bellhop's pocket (see table 10.1).

The amount paid by the guests actually totals $27: the cost of the room, plus the tip. The $29 tally double-counts the tip, but the double-counting is hidden by mathematical misdirection, in which the "total cost" paid by the guests masquerades as merely the "room cost." This riddle does not challenge the concept of summation as much as it does the way that we identify the things that need to be summed. After reading the faulty tally, the proper question to ask is not why it doesn't add to $30, but whether

Table 10.1
Proper accounting of the initial $30.

Amount	Location
$25	Cash register
$1	Guest pocket
$1	Guest pocket
$1	Guest pocket
$2	Bellhop pocket

it even should. So, a group of numbers were added; what does their tally *mean*?

10.2   A Paradox of Derivatives

Observe that the square of a variable, x^2, is obtained by summing x to itself x times (and if you don't believe that, test it on any integer):

$$x^2 = x + x + x...(x \text{ times})$$

Now, let's take the derivative of each side with respect to x:

$$2x = 1 + 1 + 1...(x \text{ times}) = 1 \times x = x$$

We have just proved that $2x = x$, or that $2 = 1$.

Claim 1
Mathematics is a human construct anyway, so certain manipulations like this one are bound to create nonsensical results. Mathematics is about picking and choosing the results that make sense.

Claim 2
The assumption that x^2 equals the sum of x to itself x times is true for natural numbers, but does not make sense for nonintegers. This causes problems when we want to take derivatives.

Discussion and Resolution
Claim 2 is correct. Not all functions are differentiable. A differentiable function must be continuous at every point in its domain. Our initial assumption was that "the square of a variable, x^2, is obtained by summing x to itself x times." This may be true of natural numbers, but it is not true of all real

numbers. How does one perform the addition operation some fractional or noninteger number of times? Can you add 2.789 to itself 2.789 times? The key assumption used to write the first line of logic is true only in a discrete sense for natural numbers, but it is not continuously true for all real numbers. Therefore, we cannot justify differentiating the right side of the first equation.

10.3 Two Equals One

Let's begin with the assumption that $a = b$, and see what trouble we can stir up:

$$a = b$$

Now multiply both sides by a to obtain

$$a^2 = ab$$

Now subtract b^2 from both sides to obtain

$$a^2 - b^2 = ab - b^2$$

Factor both sides:

$$(a - b)(a + b) = b(a - b)$$

Simplify:

$$a + b = b$$

Substitute b for a by our initial assumption:

$$2a = a$$

Simplify:

$$2 = 1$$

Now, that's not supposed to happen.

Claim 1
The variables a and b haven't been defined or pegged to any real value, so the sky's the limit. They can be used to prove anything.

Claim 2
One of the steps breaks a basic rule of mathematics.

Discussion and Resolution

Claim 2 is correct, and the rule in question is: "Thou shalt not divide by zero." Why can't you divide by zero? Take some number, like 10. Dividing 10 by zero is equivalent to asking, "What number times zero equals 10?" The answer is that there is no such number. Hence, a value divided by zero is undefined.

The first simplification stage, which eliminates $(a - b)$ from each side of the equation, represents an illegal division by zero. Since $a = b$ by assumption, $a - b = 0$, so we are not allowed to divide either side by $(a - b)$.

10.4 Summing a Divergent Series

The paradox here lies in finding the sum of the following series, called *Grandi's series*:

$$1 - 1 + 1 - 1 + 1 - 1 + 1 - 1...$$

The series could be grouped as follows:

$$(1 - 1) + (1 - 1) + (1 - 1) + (1 - 1)... = 0$$

But the series could also be grouped like this:

$$1 + (-1 + 1) + (-1 + 1) + (-1 + 1)... = 1$$

Alternatively, we could plug -1 into equation 2.1 (from chapter 2 on Zeno's Paradoxes of Motion) to obtain a result of $\frac{1}{2}$. There is something attractive about this answer, since the partial sum of the series oscillates between 1 and 0, and this answer finds their average.

Right off the bat, we have three possible but contradictory answers. The sum could be 0, 1, or $\frac{1}{2}$. Other answers are possible as well. So, what's the *right* answer?

(A method of proof that invokes the paradoxical properties of divergent series even has a name. It is known as the *Eilenberg–Mazur swindle*.)

Claim 1

Every 1 can be matched with a -1. This is clearly the most elegant matching because all such pairs are consecutive. Therefore, the one correct answer is 0.

Claim 2

Divergent series cannot be summed in the usual sense. We need new concepts—new reasonable methods of assigning a number to a series.

Discussion and Resolution

To sum a finite set of numbers, we simply perform the addition operation a finite number of times. But what if we want to sum an infinitely long series of numbers? If you tried entering the numbers into a calculator, you'd never finish. So, how could such a sum even make sense conceptually?

It makes sense if we tweak the concept of summation, as Baron Augustin-Louis Cauchy did, defining it not as the result of an infinite number of additions, but as the limit of the series of partial sums. This definition was a great innovation, whose importance was underscored by G. H. Hardy in his 1949 book *Divergent Series*:

> It does not occur to a modern mathematician that a collection of mathematical symbols should have a "meaning" until one has been assigned to it by definition. It was not a triviality even to the greatest mathematicians of the eighteenth century. They had not the habit of definition: it was not natural to them to say, in so many words, "by X we *mean* Y." ... It is broadly true to say that mathematicians before Cauchy asked not, "How shall we *define* $1 - 1 + 1 - 1 + \ldots$ but "What *is* $1 - 1 + 1 - 1 + \ldots$?" and that this habit of mind led them into unnecessary perplexities and controversies which were often really verbal.

Let's take a bit of time to understand Cauchy's concept of infinite summation and how far it takes us toward evaluating Grandi's series. First, what is meant by "partial sum?" The nth *partial sum* of a series is the sum of the first n terms. Suppose you have a series $\{a_k\}_1^\infty = \{a_1, a_2, \ldots, a_n, \ldots\}$. The nth partial sum, s_n, is defined to be

$$s_n = \sum_{i=1}^{n} a_i$$

If we were to start listing the partial sums, they would look like this:

$$s_1 = a_1$$

$$s_2 = a_1 + a_2$$

$$s_3 = a_1 + a_2 + a_3$$

$$s_4 = a_1 + a_2 + a_3 + a_4$$

$$s_5 = a_1 + a_2 + a_3 + a_4 + a_5$$

$$\vdots$$

The list of all these partial sums constitutes its own series, $\{s_k\}_1^\infty = \{s_1, s_2, \ldots, s_n, \ldots\}$. Cauchy defined the sum of a convergent series as the limit of this series of partial sums.

Let's try Cauchy's method and calculate the partial sums given by Grandi's series:

$$s_1 = 1$$

$$s_2 = 1 - 1 = 0$$

$$s_3 = 1 - 1 + 1 = 1$$

$$s_4 = 1 - 1 + 1 - 1 = 0$$

$$s_5 = 1 - 1 + 1 - 1 + 1 = 1$$

$$\vdots$$

And so it continues, yielding the following series of partial sums:

$$\{1, 0, 1, 0, 1, 0, 1, 0, \ldots\}$$

It looks like we have a problem here. This series of partial sums has no limit. It does not converge, as figure 10.1 shows. Both the original series

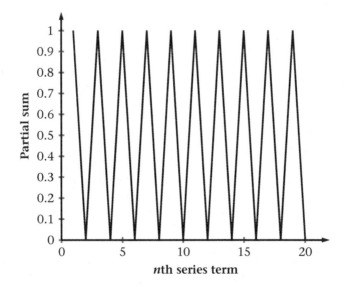

Figure 10.1
First twenty partial sums of the Grandi series.

and the series of partial sums are said to be *divergent*. Cauchy's concept of summation does not apply here, and that means we need new concepts of summation in the spirit of the correct Claim 2.

The mathematician Leonard Euler took the first steps toward developing a theory of divergent series in the eighteenth century. He wrote in his 1760 paper *De seriebus divergentibus*, as translated by E. J. Barbeau and P. J. Leah:

> Notable enough, however, are the controversies over the series $1 - 1 + 1 - 1 + 1 - 1 + \ldots$ whose sum was given by Leibniz as $\frac{1}{2}$, although others disagree. No one has yet assigned another value to that sum, and so the controversy turns on the question whether the series of this type have a certain sum. Understanding of this question is to be sought in the word "sum"; this idea, if thus conceived—namely the sum of a series is said to be that quantity to which it is brought closer as more terms of the series are taken—has relevance only for convergent series, and we should in general give up this idea of sum for divergent series. Wherefore, those who thus define a sum cannot be blamed if they claim they are unable to assign a sum to a series. On the other hand, as series in analysis arise from the expansion of fractions or irrational quantities or even transcendentals, it will in turn be permissible in calculation to substitute in place of such a series that quantity out of whose development it is produced. For this reason, if we employ this definition of sum, that is to say, the sum of a series is that quantity which generates the series, all doubts with respect to divergent series vanish and no further controversy remains on this score, in as much as this definition is applicable equally to convergent or divergent series. Accordingly, Leibniz, without any hesitation, accepted for the series $1 - 1 + 1 - 1 + 1 - 1 + \ldots$, the sum $\frac{1}{2}$, which arises out of the expansion of the fraction $\frac{1}{1+1}$, and for the series $1 - 2 + 3 - 4 + 5 - 6 + \ldots$, the sum $\frac{1}{4}$, which arises out of the expansion of the formula $\frac{1}{(1+1)^2}$. In a similar way a decision for all divergent series will be reached, where always a closed formula from whose expansion the series arises should be investigated. However, it can happen very often that this formula itself is difficult to find, as here where the author treats an exceptional example, that divergent series par excellence $1 - 1 + 2 - 6 + 24 - 120 + 720 - 5040 + \ldots$, which is Wallis' hypergeometric series, set out with alternating signs; this series, in whatever formula it finds its origin and however much this formula is valid, is seen to be determinable by only the deepest study of higher Analysis. Finally, after various attempts, the author by a wholly singular method using continued fractions found that the sum of this series is about 0.596347362123, and in this decimal fraction the error does not affect even the last digit. Then he proceeds to other similar series of wider application and he explains how to assign them a sum in the same way, where the word "sum" has that meaning which he has here established and by which all controversies are cut off.

Euler's writing shows that he believed mathematics would benefit from having a systematic way of associating numbers with divergent series. His ideas on the subject had been brewing for years. His 1745 letters to Christian Goldbach and Nicholas Bernoulli reveal that he believed every series should be assigned a certain value, but that such values should not be denoted as "sums," as sums imply true addition. Euler's work in this area took a temporary toll on his reputation, as many nineteenth-century mathematicians banished divergent series from mathematical study. People misunderstood his ideas and began thinking of him as a careless thinker. He is appreciated now not as careless, but as one of the greatest mathematical minds ever, whose theory of divergent series proved to be well ahead of his time.

The task at hand, in the spirit of Euler's theory, is not really to add a group of numbers, but to assign a finite numerical value to a divergent series in some reasonable way. Mathematicians have introduced a variety of ways to do this. They have also introduced criteria for what constitutes a reasonable method. The methods do not all give the same results or meet the same criteria, so it is important for mathematicians to specify the method they are using when they "sum" a divergent series. These sums are known as *regularized sums*, and methods of calculating them fall under a category of processes known as *regularization*.

Let's talk first about the criteria of regularization. What makes a method reasonable? There are three important elements to check.

The first criterion is *regularity*. A summation method A() is said to be regular if, when applied to any convergent series, it returns Cauchy's correct answer: the limiting value of the partial sums. Why is this a reasonable criterion? Cauchy's method is a generalization of the concept of "addition"; it allows us to "add" in ways that wouldn't be possible with a mere calculator. Regularization methods generalize the concept even further; they assign finite values to "sums" that are incalculable even with Cauchy's method. Each generalization is meant to extend the concept of "addition" so that new, more challenging types of inputs can be handled, but not change the prior, more restrictive concepts of "addition." In other words, we want to keep all the same outputs to our old inputs. At the same time, we want to assign outputs to new kinds of inputs that couldn't be handled before.

The second criterion is *linearity*. Let a and b each be a series, and let c be some constant scalar, real or complex. A summation method A() is said to be linear if $A(ca + b) = cA(a) + A(b)$.

The third criterion is *stability*. Suppose you have a series a_k, which is shorthand for $\{a_k\}_1^\infty$, and suppose you have a summation method A(), such that

$$\sum_{n=1}^{\infty} a_n = A(a_k)$$

The summation method A() is said to be stable if

$$\sum_{n=2}^{\infty} a_n = A(a_k) - a_1$$

In other words, stability is achieved if the summation method subtracts from the result any leading terms that are omitted. The second and third criteria align with our most basic notions of how addition should work.

Two summation methods will now be introduced. Mathematicians have developed many methods of regularization beyond these two summation methods. The idea here is to show some more intuitive examples of how it can be done.

Cesàro Summation

This summation method was named for Ernesto Cesàro, who formalized its definition, although the method had been used before him. Suppose you have a series a_k, again shorthand for $\{a_k\}_1^\infty$. Let the series of partial sums be s_k. The Cesàro summation method outputs the limiting value of the *average* of the partial sums. Formally,

$$A(a_k) = \frac{1}{n} \lim_{n \to \infty} \sum_{i=1}^{n} s_i$$

Let's use this method to evaluate Grandi's series. We found before that the first five partial sum values are $\{s_1, s_2, s_3, s_4, s_5\} = \{1, 0, 1, 0, 1\}$.

The average of the first partial sum is itself: 1 divided by 1. The average of the first two partial sums is $\frac{1+0}{2} = \frac{1}{2}$. The average of the first three partial sums is $\frac{1+0+1}{3} = \frac{2}{3}$. The average of the first four partial sums is $\frac{1+0+1+0}{4} = \frac{1}{2}$. The average of the first five partial sums is $\frac{1+0+1+0+1}{5} = \frac{3}{5}$.

The series generated by taking average partial sums is $\{1, \frac{1}{2}, \frac{2}{3}, \frac{1}{2}, \frac{3}{5}, ...\}$. This series converges to a value of $\frac{1}{2}$, which is the Cesàro sum of the Grandi series (see figure 10.2).

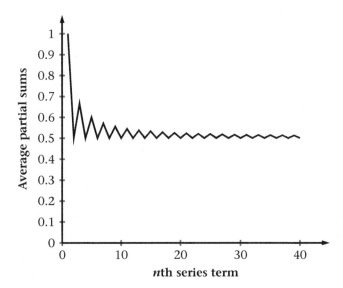

Figure 10.2
The average partial sums of the Grandi series converging to $\frac{1}{2}$.

This method of summation can handle oscillating sequences by smoothing them out, but if the partial sums of a series tend toward infinity, then so do the averages of those partial sums. In such cases, this method does not work.

Abel Summation

The Norwegian math prodigy Niels Henrik Abel famously said: "Divergent series are the invention of the devil, and it is shameful to base on them any demonstration whatsoever." Despite his apparent antipathy for the subject, Abel nonetheless developed a summation method consistent with (and even more powerful than) Cesàro summation. It can handle any series that Cesàro summation can, and more besides; and when they work, they produce the same answer.

Formally, for some series a_k whose initial term is indexed by zero (a_0, instead of a_1 as used previously in this chapter), the Abel summation function is defined as follows, where such a limit exists:

$$A(a_k) = \lim_{z \to 1^-} \sum_{i=0}^{\infty} a_i z^i$$

The minus sign in the limit indicates that the limit is taken from the left; that is, as z approaches 1 from below.

This method is best understood through example. Let a_k be the Grandi series. The summation on the right side of the Abel equation becomes

$$\sum_{i=0}^{\infty} a_i z^i = 1 - z + z^2 - z^3 + z^4 - z^5 \dots$$

Using the logic used to derive equation 2.1, it can be shown that

$$1 - z + z^2 - z^3 + z^4 - z^5 \dots = \frac{1}{1+z}$$

for any $z \in (-1, 1)$. Returning to the original Abel equation, we must take the limit:

$$A(a_k) = \lim_{z \to 1^-} \left(\frac{1}{1+z} \right) = \frac{1}{2}$$

The limit exists. Hence, the Abel summation of Grandi's series is $\frac{1}{2}$.

Intuitively, the idea of Abel summation is to extend the logic of equation 2.1, which only works for $|z| < 1$, to the case where $z = 1$ by taking a limit from below.

10.5   Summing the Naturals

We are now going to examine the infinite series whose terms are the natural numbers:

$$1 + 2 + 3 + 4 + 5 \dots$$

The partial sums are $\{1, 3, 6, 10, 15, \dots\}$, diverging to infinity. Try using Cesàro summation or Abel summation, and you will find that those methods do not work. In fact, any method that is both stable and linear will not work. Intuitively, it seems like we shouldn't need formal methods. Why should we? With 1, plus 2, plus 3, and so on, each time, we're adding something *larger*, so certainly the only value that can be assigned to this series is infinity.

Or is it?

In fact, a number of methods can be used to assign a real value to the series. That value, believe it or not, is $-\frac{1}{12}$.

Yes, not only a fraction, but a *negative* fraction, assigned to the series whose elements are the positive integers.

Claim 1

Impossible. This series clearly diverges to positive infinity. There is no way to assign a real, finite value to it, let alone a negative, fractional value.

Claim 2

Recalling that we are not "summing" in the traditional sense, and remaining open to considering logically defensible ways of assigning values to divergent series, we will find $-\frac{1}{12}$ to be a reasonable choice.

Discussion and Resolution

Claim 2 is correct. One such method will be presented, but there are a variety of others. To understand the method, a few new mathematical concepts must be introduced.

Let's begin with a review of complex numbers. A *complex number* is a number that has both a real and an imaginary component. For example, $2 + 3i$ is a complex number whose real part is 2 and whose imaginary part is $3i$, where $i = \sqrt{-1}$. A *complex function* is a function whose range is in the complex numbers.

Calculus can be extended to complex functions; we can take derivatives of complex functions at certain points if they are complex differentiable at those points. We won't go into the exact definition of *complex differentiable*, but the same intuition applies as with differentiable real-valued functions: There should be no breaks or kinks. If a complex function is complex differentiable at every point in a region, then it is said to be *analytic* in that region.

Complex functions may be limited in their domains. There may be values for which a function makes sense, and other values for which it doesn't. By a process known as *analytic continuation*, mathematicians can extend the domain over which a complex function is defined. The only condition that a continued function must satisfy is that it must be complex differentiable everywhere, including in the extended domain.

Let's take a relevant example. Consider the following function:

$$f(s) = \sum_{n=1}^{\infty} \frac{1}{n^s}$$

where s is complex valued. (In standard notation, $s = \sigma + it$, where σ is the real part and it is the imaginary part; s is a real number if $t = 0$.) Let's plug in a value of $s = 2$ to get a sense of what this looks like in expanded form:

$$f(2) = \frac{1}{1^2} + \frac{1}{2^2} + \frac{1}{3^2} + \frac{1}{4^2} + \frac{1}{5^2}\ldots \tag{10.1}$$

It can be shown that the output series of the function $f(s)$ will converge to a finite value, so long as the real component σ in the input s has a value greater than 1. But if the real component is equal to or less than 1, the output series will not converge. For example, let's try the value -1:

$$f(-1) = \frac{1}{1^{-1}} + \frac{1}{2^{-1}} + \frac{1}{3^{-1}} + \frac{1}{4^{-1}} + \frac{1}{5^{-1}}\ldots = 1 + 2 + 3 + 4 + 5\ldots \tag{10.2}$$

By plugging in an input value of -1, we have generated a series of natural numbers! This will become useful very soon.

The function f is analytic (complex differentiable), so long as $\sigma > 1$. But the function does not really make any sense in its current form for any input value with $\sigma \leq 1$ because the output series diverges to infinity for any such value.

This function is, therefore, a perfect candidate for analytic continuation. Fortunately, the condition that the analytically continued function be complex differentiable everywhere is so restrictive that it locks in exactly *one* answer. There is only one way to analytically continue the function f.

The analytically continued form of f is one of the more famous functions in mathematics. It is known as the *Riemann zeta function*, named after Bernhard Riemann, and is expressed as $\zeta(s)$.

Mathematicians don't know everything there is to know about the Riemann zeta function. A \$1 million prize awaits anyone who can prove or disprove the Riemann Hypothesis, which conjectures as to the values at which the function equals zero. Proving or disproving the Riemann Hypothesis is one of the seven Millennium Problems, considered among the most difficult challenges in mathematics, of which only one has been solved since they were stated in 2000.

We do know, however, that the function has some remarkable properties, some of which are so remarkable they seem mystical until they are explained. For instance, in his solution to the "Basel problem," Euler calculated the value of equation 10.1, or $\zeta(2)$, to be exactly $\frac{\pi^2}{6}$.

But perhaps one of the most fascinating output values of the Riemann zeta function is that at -1, which equation 10.2 relates to our original object of study: the series of natural numbers.

It turns out that $\zeta(-1)$ can be shown to equal exactly $-\frac{1}{12}$.

Is it true that $1+2+3+4+5\ldots=-\frac{1}{12}$? Not in the traditional sense. But the two are deeply associated, and you have now seen one reasonable method by which to make the association. There are other methods which suggest the same association. Readers may be interested in independently learning about one developed by Srinivasa Ramanujan, which is covered extensively in mathematical literature.

11   Classical Physics

Nicholas Laurita

Truth is ever to be found in simplicity, and not in the multiplicity and confusion of things.

—Isaac Newton

As you will soon discover, the physics of the most extreme scales of the universe, of the subatomic world of quantum mechanics and the hyperspeed realm of special relativity, give rise to remarkable paradoxes. Yet these extreme contexts feel far removed from the world of everyday experience. We need not travel quite so fast, or shrink quite so small, to discover interesting paradoxes found in classical mechanics and thermodynamics—many of which have practical implications for human existence. Several of these paradoxes are directly testable in laboratory conditions and continue to garner modern scientific interest. This chapter tackles a few of these paradoxes, beginning with a brief introduction that assumes that the reader is familiar with the very basics of classical mechanics (objects respond to forces in accordance with Newton's laws of motion). Fundamental concepts of thermodynamics are the focus here.

Thermodynamics governs much of how we experience the world. Its laws form the physical foundation that allows our refrigerators to keep our food cold, our thermoses to keep our coffee hot, and the engines in our cars to extract useful energy from gasoline. Thermodynamics also explains some less obvious aspects of our lives. Have you ever wondered why heat flows only from hot objects to cold objects and never in the reverse direction? Or why houses tend to get messier over time instead of cleaner? Why do so many phenomena naturally seem to occur in one direction, but not the other?

It will be useful to make a distinction between the philosophies of thermodynamics and classical mechanics. In classical mechanics, we are given a set of initial conditions of an object at a certain beginning time. For instance, we may know an object's position, velocity, or momentum, and then the goal is to find the value of these or other variables at a later time using Newton's laws of motion.

Let's take a simple example. Picture a ball about to roll down a hill. In classical mechanics, we may ask: Given the height of the hill, what will the velocity of the ball be once it reaches the bottom? Or perhaps, given the height of the hill and the mass of the ball, what will the ball's momentum be at the bottom? Now, it gets trickier when we want to generalize these concepts to a container of gas in which every individual gas molecule has a position, velocity, and momentum. With trillions upon trillions of gas molecules in even a tiny volume of gas, there is no way we could possibly keep track of all these variables, even if we could measure them, to make an accurate prediction of the state of the gas at a later time. This is the fundamental difference in philosophy that thermodynamics introduces. Instead of focusing on variables related to individual molecules of a gas, thermodynamics focuses on aggregate, or macroscopic, variables of the entire volume of gas. In thermodynamics, we would not evaluate the individual velocity of each gas molecule, but rather the temperature of the entire gas. Other questions might include: How much internal energy does the entire gas have? What is the pressure of the entire gas? Defined formally, thermodynamics is the study of heat and energy transfer between systems.

We can delve deeper by introducing the laws of thermodynamics, a set of four principles that provide a foundation for the field. If our goal is to study heat transfer between systems, then we must first define the concept of temperature, which is the objective of the so-called zeroth law of thermodynamics. (The name stems from the fact that it was added after the first three laws were formalized.) The zeroth law states: "If two systems are each in thermal equilibrium with a third system, then they are also in thermal equilibrium with each other."

This is a fairly intuitive concept. Our experience tell us that heat flows from hot to cold objects until both reach the same intermediate temperature. At this point, heat transfer between the objects stops, and *thermal equilibrium* has been reached. The zeroth law serves as an empirical definition of temperature. It ensures that if we measured the temperatures of

Systems A and B using a thermometer and found them to be equal, then Systems A and B would be in thermal equilibrium with each other when placed in thermal contact.

But what are we really measuring when we gauge the temperature of something? At any finite temperature, each molecule of a substance has movement that can be decomposed into *translations* (movement from one place to another), *rotations*, and *vibrations*. The atoms in solids have zero translational movement, but instead rotate and vibrate about their equilibrium positions. Molecules in fluids (liquids and gases) have all three kinds of movement. Any molecule that is moving has kinetic energy. *Temperature* is the average kinetic energy of the molecules in a substance. (The concept of temperature has to be fine-tuned as we look closely at certain materials and states of matter; this is an introductory concept of temperature.)

In 1827, the botanist Robert Brown directly observed some effects of molecular movement. Looking through a microscope, he watched as tiny grains of pollen moved randomly through the water in which they were suspended. We understand now that the pollen grains were being pushed around due to collisions with the molecules of water. Brown's observation gave rise to the term *Brownian motion*, which describes the random movement of particles suspended in fluid.

If we could zoom in and observe the individual molecules of a gas, we would find them randomly moving and colliding with each other, exchanging energy in the process. A measurement of the velocities of the molecules at any one time would reveal a distribution of velocities. Some molecules would be traveling faster or slower than average, but most would be traveling close to average speed. A snapshot at a later time would reveal a similar distribution of velocities, but now some of the previously slow molecules, having since collided with faster ones, would have picked up energy and speed—and faster ones will have lost energy and speed. Although some individual molecules would have changed their velocities, the *average* velocity of all the molecules would have remained the same.

Hotter substances have molecules that are moving faster, on average, than colder ones. In classical thermodynamics, thermal motion is always present, slowing as we cool to lower temperatures, all the way down to absolute zero, the coldest possible temperature, at which point all motion stops. (Quantum mechanics predicts the continuation of some motion even at absolute zero, but this distinction is beyond the scope of this introduction.)

With temperature defined, we can state the first law of thermodynamics, which ensures energy conservation: "The total energy of a closed thermodynamic system is constant and given by the amount of heat supplied to the system, minus the amount of work done by the system on its surroundings." A system is "closed" if only energy, not matter, can enter or leave the system. Picture a container of gas that can be heated, but that allows no molecules to penetrate its walls. The first law can be written mathematically as

$$\Delta U_{\text{int}} = Q - W \tag{11.1}$$

Here, ΔU_{int} is the change in internal energy, Q is the amount of transferred heat, and W is the work done by the system. The first law can be understood qualitatively without too much detail. The internal energy of a system describes the state of the system; it results from the the cumulative contributions of all the kinetic energies and potential energies of every constituent molecule. We can change the internal energy of the system by transferring energy to or from the system as represented on the the right side of equation 11.1. We are most familiar with adding or removing heat. For example, refrigerators remove heat from food to keep it fresh. The first law says that this removed heat must go somewhere else (to the air in our kitchens). The internal energy of a system can also be altered by performing work. Take the gasoline engines in our cars, for example. The pistons of the engine compress fuel before ignition. This high-pressure gas then pushes the pistons back down, turning the motor in the process, thereby creating mechanical work from the internal energy of the gas. The first law ensures that no energy is gained or lost in this process or other thermodynamic processes.

The second law of thermodynamics is perhaps the most consequential of the four laws, but also likely the most difficult to understand. The second law states: "The entropy of the universe can never decrease over time." But what is *entropy*? The *entropy* of a system is a measure of the number of microscopic arrangements, or *microstates*, that result in the same macroscopic variables, or *macrostates*. We can define entropy mathematically as

$$S = \ln(\Omega) \tag{11.2}$$

where S is entropy and Ω is the number of microstates that result in the same macrostate. (This is a dimensionless definition of entropy, whereas conventionally, entropy has the dimension of energy per temperature.)

To better understand entropy, let's visit a familiar context: flipping coins. Assume that all coins are perfectly fair; that is, the probability of heads or tails is $P(H) = P(T) = 0.5$. Suppose we are tasked with calculating the expected number of heads obtained by flipping ten coins. We approach this task by asking how many different combinations of the flipped coins result in a given number of heads.

There is only one arrangement that produces zero heads: Each coin must come up tails. How about one heads? In this scenario, only one coin must show heads after all the coins have been flipped. Since we have ten coins, there are ten ways that we can end up with one coin showing heads. If we continue to two heads, the number of available arrangements (*combinations*) increases to 45. Three heads has 120 combinations, four heads has 210 combinations, and five heads has 252 combinations. For a larger number of heads, the number of combinations start to decrease again.

In figure 11.1, the left panel shows the number of possible combinations (microstates) that result in a given number of heads (macrostate) after ten coins have been flipped. The right panel displays the corresponding entropy calculated by equation 11.2. Not only is five heads the most probable outcome, but it is also the macrostate with the highest entropy.

This simple example provides insight into the nature of thermodynamic systems. Entropy is a measure of how many microscopic arrangements of particles result in the same macroscopic variables, such as temperature or pressure. Systems reach equilibrium by settling into the macrostate with the greatest number of microstates—that is, with the highest entropy.

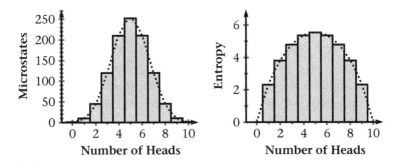

Figure 11.1
Microstates and entropy in a coin-flipping experiment.

In a sense, entropy is measure of the disorder of a system. Take our coin-flipping example again. The macrostates with either zero or ten heads are the most ordered states, achievable only if all the coins show the same result. However, they are also the macrostates that are the least likely to occur, and they possess the lowest entropy. The most likely outcome is the most disordered state, five heads and five tails, which is also the macrostate with the most entropy. The second law of thermodynamics ensures that systems tend toward the state with the most entropy over time. The entropy of a closed system can never decrease; systems must tend toward disorder until equilibrium is reached, at which point the change in entropy over time becomes zero.

The second law also dictates which processes are spontaneous, or happen naturally, as opposed to other processes that require expending outside energy. Imagine, for example, putting a droplet of food coloring into a glass of water. On its own, the pigment will slowly become uniformly distributed throughout the water. The pigment does not reassemble itself into the original droplet. Why not? Ultimately, such a reassembly would be incredibly *unlikely*. There are far more microstates that support a "uniformly colored" macrostate than there are microstates that support a "droplet" macrostate. When the droplet first enters the water, it's in a highly ordered state. The water and the droplet are distinct. The second law implies that the system will evolve in a way that increases its entropy and disorder. Here, that means spreading the pigment uniformly throughout the water. Once completely dispersed, the system has reached equilibrium, and the change in entropy over time is zero. The food coloring does not reassemble itself into a droplet because such a reassembly would represent a decrease in net entropy. Separating the pigment from the water after dispersion would be possible only by expending energy. In fact, the second law ensures that we would have to create more entropy in the universe by reassembling the pigment into its original droplet than we would save. Processes whereby entropy increases over time are said to be *irreversible*, because reversing them without outside energy would violate the second law of thermodynamics.

The irreversibility of processes explains why events have preferred directions in time. Heat flows from hot objects to cold objects, not the other way around. Chewing gum loses flavor with each bite; it doesn't become more flavorful. Homes get messier without effort, and they never seem to clean themselves. (Perhaps the most unbelievable of Mary Poppins's powers is her ability to decrease entropy in the universe.) All these processes are dictated by the second law of thermodynamics, which favors disorder in time.

The third and final law of thermodynamics is less relevant for our paradoxes but is stated here for the sake of completeness: "The entropy of a perfect crystal is zero at absolute zero." This follows from our previous discussion. At absolute zero, the lowest possible temperature, the classical motion of all atoms in a crystal ceases, and the crystal becomes uniform everywhere, achieving the most ordered state possible. This singular arrangement of atoms in a crystal provides a reference point for measuring entropy; it serves as a state of zero entropy, against which entropy can be measured at higher finite temperatures.

With the laws of thermodynamics in hand, we have one final thing to discuss before getting to the paradoxes. That is the concept of *perpetual motion machines*. The idea of a perpetual motion machine is a surprisingly old one, dating back to the Middle Ages. The idea is fairly simple: Can a machine be developed that performs more work than the energy input into the machine? Or perform work with no input energy at all?

We should be skeptical. We know that (nonelectric) cars need gasoline to run, and electronics need to be charged or given batteries to operate. Daily experience points to the need for an input energy source if work is to be performed. This is consistent with our discussion of the first law of thermodynamics. Equation 11.1 shows that the work created in a thermodynamic system is at most equal to the internal energy of the system, and even this is possible only in the ideal case, in which no heat gets created. But how applicable is the first law at scales different from those of our everyday experiences? Does it apply at a microscopic level that considers individual gas molecules? Is there some clever machine that would allow the laws of thermodynamics to be subverted? The first few paradoxes given here attempt to do just this.

11.1 Maxwell's Demon

Maxwell's demon is a well-known paradox in physics originally conceived by physicist James Clerk Maxwell, who is perhaps best known as the father of electromagnetism. He was also a pioneer in the early days of thermodynamics and introduced this paradox publicly in his 1872 book *Theory of Heat*. The "demon" in the paradox was not in his original description; it was

Figure 11.2
Maxwell's demon selectively opening the door.

added in a 1874 journal manuscript by Lord Kelvin. Although the paradox
is believed to have been resolved, more modern twists, with the aid of rapid
advances in nanotechnology, have led to continued interest in Maxwell's
demon to this day.

In its simplest form, the paradox is as follows: Consider a container of
gas at a given temperature, T. A microscopic demon sits in the middle of
the container, guarding a trapdoor between two chambers of the container,
which have equal volume. (See figure 11.2.) Assume that the door has neg-
ligible weight—no need to worry about the work expended by moving the
door. The demon observes individual gas molecules on either side of the
door, only opening the door to allow slower (colder) molecules to move into
the right chamber and faster (hotter) molecules to move into the left chamber. He does this in a manner that maintains constant pressure throughout
the entire system.

Over time, separating the molecules in this fashion creates a temperature
difference, ΔT, between the two chambers. Theoretically, this difference
could be used to perform usable mechanical work by a heat engine. Thus,

work has been generated from nothing but the ambient properties of the gas. No additional energy has been provided. This is a form of perpetual motion machine.

Claim 1

Such a partitioning of hot and cold molecules is scientifically valid despite the apparent reduction in entropy of the gas and the fact that heat appears to flow from the initial colder gas into the hotter left chamber. Therefore, the second law of thermodynamics must be flawed, or at least it applies differently on the microscopic scale.

Claim 2

Sorting the gas molecules into hot and cold chambers does not violate the second law of thermodynamics. Somewhere in the process, there must be additional entropy created which must be greater than or equal to the entropy reduction derived from sorting the gas molecules.

Discussion and Resolution

Claim 2 is correct. Although an initial qualitative resolution was provided by Marian Smoluchowski in 1912, the first quantitative approach to the problem appeared in a 1929 paper by physicist Leó Szilárd.

Szilárd argued that the paradox overlooks the entropy created by the demon when making a measurement of a molecule's velocity before determining if the door should be opened. The mere act of measurement is itself irreversible, and therefore creates entropy. The demon and the gas container must be treated together as "the system." The entropy created by performing the measurements is in fact greater than the entropy reduction caused by separating the molecules into hot and cold chambers. The total entropy of the gas-demon system has increased during the sorting process, in accordance with the second law of thermodynamics.

This measurement-entropy argument was taken further in papers by Leon Brillouin and Denis Gabor, who quantitatively demonstrated that if the demon uses light to perform a measurement of a particle's velocity (a reasonable technique), the entropy of the demon-gas system would increase—even if the demon did so at the maximum possible efficiency.

For the following thirty years, Maxwell's demon appeared to be fully resolved. In 1982, however, physicist and information theorist Charles

Bennett demonstrated that a measurement of a molecule's velocity could be done so reversibly, with no entropy created in the measurement process. Bennett argued that it isn't the measurement process itself that creates entropy, but rather the erasure of information after the measurement is made in preparation for the next measurement. Bennett's argument built upon the previous work of physicist and IBM researcher Rolf Landauer, who showed in the 1960s that erasing information is itself an irreversible process and costs at least a minimum amount of entropy. Today, this concept is known as *Landauer's principle*.

Suppose the demon is a simple computer that can store one bit of information at a time. That bit encodes the velocity of a single molecule as "fast" or "slow." The computer observes a molecule approaching the door and makes a reversible, or entropy-free, measurement of its velocity, opening the door or not depending on the result. The computer must now erase this memory before the next molecule approaches, reducing the memory state from either "fast" or "slow" to an empty slate of "no measurement has been made." This is an irreversible process; once the bit has been erased, it is gone and cannot be retrieved. Irreversible processes correspond to a net increase in entropy. In real computers, this manifests as a slight increase in temperature of the computer after erasure. Bennett demonstrated that the entropy created by erasing data is at least enough to compensate for any entropy reduction that results from the demon's sorting of the molecules. So the total entropy of the gas-demon system increases, all things considered.

This is a fascinating insight with real and verified implications. We have all noticed that our computers feel hot after prolonged use. Most of this heat results from currents running through restive components in our computers, but a small amount results from erasing data! It's a result of the universe's tendency toward disorder. Attempt to bring greater order to your computer, and the universe fights back.

But wait a minute. Couldn't the demon just store more than one bit of information at a time? Couldn't the demon store a large array of bits? He can, but any physically realizable system will have finite memory space. Eventually, the demon will have to start erasing old measurements to make room for new ones. In the end, the second law of thermodynamics still wins.

11.2 Brownian Ratchet

The Brownian Ratchet Paradox, which is sometimes called the *Feynman-Smoluchowski Ratchet Paradox*, was first discussed in 1900 by physicist and inventor Gabriel Lippmann, who would later win the Nobel Prize in Physics for inventing color photography. The paradox garnered popular interest after its inclusion in Richard Feynman's lecture series *The Feynman Lectures on Physics*.

The paradox centers around the device shown in figure 11.3. A paddle wheel, which sits in a chamber of gas at temperature T_1, is connected by a shaft to a gear that sits in a similar container at temperature T_2. A pawl, which pushes against the teeth of the gear, ensures that the gear can rotate only in one direction. (The pawl and gear together constitute the ratchet.) The entire system is built microscopically, so collisions with individual molecules of the gas are strong enough to turn the paddle wheel and rotate the gear. Normally, random collision between the gas molecules and the paddle wheel would force the gear to rotate with equal probability in both directions. However, thanks to the pawl, the gear is restricted to turn in only one direction, which can be used to perform work. For instance, the device can be used to lift a mass on a string, as shown in the diagram. Once again, it appears that the second law of thermodynamics has been cheated! This device seems to create usable work from the ambient properties of the gas.

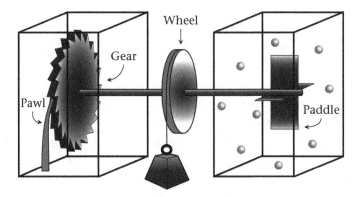

Figure 11.3
The Brownian ratchet mechanism.

Claim 1

The Brownian ratchet is a physically realizable form of a perpetual motion machine. Maxwell's demon may not have been clever enough to beat the second law of thermodynamics, but now we have.

Claim 2

The device does not violate the second law of thermodynamics. Somewhere in the design, there must be a flaw that prevents its operation as described.

Discussion and Resolution

Claim 2 is correct. The Brownian ratchet actually can be thought of as an example of Maxwell's demon. It is reminiscent of a thought experiment considered by Smoluchowski in 1912, wherein the demon is replaced with a one-way door.

Smoluchowski argued that the pawl will undergo thermal vibrations of its own because it is at finite temperature. These thermal vibrations will cause the pawl to bounce up and down, allowing the gear to slip backward each time the pawl momentarily jiggles away. Remember that the whole system is microscopic and highly susceptible to even tiny vibrations.

Feynman showed that if the ratchet and paddle wheel are at the same temperature ($T_1 = T_2$), then the gear will slip backward as often as it rotates forward, resulting in no average rotation of the gear over time. Only if the ratchet is at a different temperature than the paddle wheel ($T_1 > T_2$), can we expect the wheel to rotate more in one direction than the other. This rotation derives from the temperature difference; it is a miniature heat engine that does not violate the second law of thermodynamics. In time, the temperatures between the two chambers would equilibrate, at which point no further work could be extracted in this manner.

What if we make the whole mechanism larger, so the pawl is less susceptible to vibrations? There's a trade-off. A larger gear would also be harder to rotate. Make the device too big, and the colliding gas molecules will not be strong enough to turn the paddle wheel. There is no regime such that the device is large enough to overcome the pawl's thermal vibrations and yet small enough for the paddle wheel to turn. The second law of thermodynamics still prevails. No matter how cleverly designed, no machine can achieve "perpetual motion" status. Any fix to any problem simply creates another problem. Perhaps Leonardo da Vinci said it best:

Oh, ye seekers after perpetual motion, how many vain chimeras have you pursued? Go and take your place with the alchemists.

11.3 Feynman Sprinkler

The Feynman Sprinkler Paradox originated with physicist Ernst Mach, who suggested the paradox in his 1883 book *The Science of Mechanics*. The paradox earned its contemporary moniker after Richard Feynman recalled an attempt to build the device in his autobiography, *Surely You're Joking, Mr. Feynman!* He had done so while a graduate student at Princeton University.

The concept of the paradox is quite simple. A typical sprinkler consists of a freely moving wheel from which water is expelled in jets. The angle of the emitted water ensures that a force is exerted on the wheel, making the wheel rotate as it operates, as shown in figure 11.4. The Feynman Sprinkler Paradox asks: Assuming that we ignore the effects of friction and the viscosity of the fluid, what happens if we run the sprinkler in reverse? In other words, if we submerge a sprinkler and force it to suck water in instead of shooting water out, does it rotate in the opposite direction—*toward* the incoming fluid (see figure 11.4)?

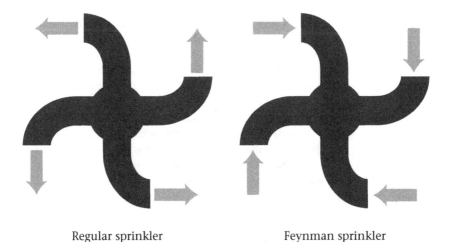

Regular sprinkler   Feynman sprinkler

Figure 11.4
Regular sprinkler versus the Feynman sprinkler.

Claim 1

A reverse sprinkler experiences a force in the opposite direction as it would if operated normally. Therefore, the reverse sprinkler should rotate in the opposite direction, toward the incoming fluid.

Claim 2

A reverse sprinkler experiences a force in the same direction as it would if operated normally. Therefore, the reverse sprinkler should rotate in the same direction, away from the incoming fluid.

Claim 3

A reverse sprinkler experiences no net forces at all and therefore does not rotate.

Discussion and Resolution

Claim 3 is correct, albeit with some caveats. Feynman never discussed the results of his experiment while at Princeton and also never provided a solution to the problem. However, Edward Creutz, who was in charge of the Princeton Cyclotron Laboratory at the time, later revealed that he had assisted Feynman in building and testing the device. He recalled the experiment in 2005 as follows:

> We watched the sprinkler as the pressure built up, forcing water into the top of the sprinkler, and in a few minutes the sprinkler head turned about 5 degrees in the direction a sprinkler usually turns when water is forced into it through its base. There was a little tremor, as [Feynman] called it, and the sprinkler head rapidly moved back to its original position and stayed there. The water flow continued with the sprinkler stationary. We adjusted the pressure to increase the water flow, about five separate times, and the sprinkler did not move, although water was flowing freely through it in the backwards direction ... The carboy then exploded, due to the internal pressure. A janitor then appeared and helped me clean up the shattered glass and mop up the water. I don't know what [Feynman] had expected to happen, but my vague thoughts of a time-reversal phenomenon were as shattered as the carboy.

The Feynman sprinkler does not rotate. But why not? The solution, it turns out, is quite nuanced, and it has been a source of ongoing debate. A thorough treatment of the problem, and one that is accessible to nonphysicists, was provided in 2004 by Alejandro Jenkins and further explained in a 2011 manuscript. The Jenkins solution is outlined here. For simplicity, we'll

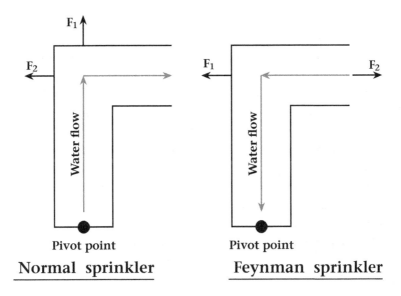

Forces acting on sprinklers during a steady state of water flow.

assume (as Jenkins did) that the sprinklers are shaped with right angles, as shown in figure 11.5, and we can imagine that both sprinklers are totally submerged in water.

In order to understand why the Feynman sprinkler does not rotate, let's first try to understand why a normally operated sprinkler does rotate. Let us consider the forces acting on the sprinkler, of which there are two. The first force results from the water rushing through the sprinkler head and hitting the top inside wall of the sprinkler. Let's refer to this force as F_1, as shown in figure 11.5.

The second force results from the pressure difference inside and outside the sprinkler head. The pressure inside the sprinkler head must be higher than the pressure outside; otherwise, the water wouldn't exit the sprinkler. This pressure difference results in a force (pressure equals force divided by area). Let's say that the water leaving the sprinkler does so with force F_2. Then, by Newton's third law of motion, we know that the sprinkler head must experience a force that is both equal and opposite to F_2. This is the same principle that allows a rocket to lift off the ground. A rocket fires high-pressure gas from its thrusters, experiencing an equal but opposite force propelling it upward.

With the forces identified, we can now understand why the regular sprinkler rotates. Forces F_1 and F_2 are orthogonal. The sprinkler cannot move in the direction of F_1, but F_2 causes it to rotate away from the outflowing water.

Let's identify the forces acting on the Feynman sprinkler. Here, the pressure inside the sprinkler must be less than that of the surrounding fluid; otherwise, the water would not enter the sprinkler. The pressure difference produces a force F_2 acting on the sprinkler in the direction, as shown in figure 11.5. There is also a second force acting on the sprinkler: F_1, caused by the impact of the water as it enters the sprinkler and hits the back inside wall of the sprinkler head. Observe that F_1 and F_2 are no longer perpendicular, but opposite in direction.

Jenkins demonstrates that these two forces perfectly offset each another. The Feynman sprinkler will recoil briefly as the first bit of water enters the head when turned on; and it will recoil briefly when turned off. But during a steady state of water inflow, the two opposing forces act in balance, preventing the Feynman sprinkler from rotating.

This analysis holds for the ideal case that ignores the effects of friction and viscosity. Practically, the viscosity of the fluid will cause F_2 to outweigh F_1 slightly, disturbing the balance and causing the Feynman sprinkler to rotate slowly toward the incoming fluid. However, friction between the sprinkler head and its base at the pivot point will tend to work against the effects of viscosity, making any rotation more difficult to achieve. Whether a real Feynman sprinkler rotates at all depends on the specifics of the design and the properties of the fluid. But no rotation occurs in an ideal system during a steady state.

12   Special Relativity

Aidan Chatwin-Davies

Einstein, in the special theory of relativity, proved that different observers, in different states of motion, see different realities.
—Leonard Susskind

Relativity and quantum mechanics together defined physics in the twentieth century. Nearly every major development in physics since has relied on one or both of these two theories. Yet, despite their prevalence in modern physics, we get little hands-on experience with relativity and quantum mechanics in our daily lives. It's therefore easy to trick ourselves into finding apparent contradictions as we think about the implications of these two theories, without the intuition provided by our senses.

This chapter focuses on paradoxes in *special relativity*, a theory that Albert Einstein developed at the beginning of the twentieth century and that he later generalized to the theory of general relativity. Special relativity is based on two simple, natural, and believable principles from which complicated, unusual, and surprising conclusions follow.

The first principle is that there is no absolute notion of being at rest. All motion is relative. Have you ever been in a car (or train) as another vehicle (or train) pulled up alongside you, and had trouble deciding whether you or the other vehicle were moving? If you stripped away all points of reference—say, you and the other vehicle were out in empty space, with no stars in sight—could you even decide whether you or the other vehicle were moving? The principle at hand asserts that the question is unanswerable. You are just as correct in thinking that the other vehicle is moving while you remain stationary as the passengers of the other vehicle are in thinking that *they* are stationary while you are moving. The answerable question is whether you and the other vehicle are moving *in relation to each other*.

This principle is not without its limitations, however. If you felt a push as your car sped up, or the train jostled you due to bumps in the rails, then you could reliably conclude that you are, in fact, moving, even if you stripped away all reference points. The first principle only applies to observers (physics lingo for people, detectors, or even inanimate objects—anything with respect to which observations can be made) that move uniformly with constant speed and have no forces acting on them as they move. We call such motion *inertial motion* and can state the first principle more precisely as follows:

Principle I
Given any two observers in uniform inertial relative motion, either is equally entitled to consider him/herself at rest.

Since it will be useful later, let's also introduce the concept of a *reference frame*. Without going into mathematical details, take a set of three axes going out from your location in the directions of the three dimensions of space—x, y, and z. These axes, which you use to measure distances and positions, together with a clock which you use to measure times, make up your abstract reference frame, with you sitting at the *origin*. Another observer will have another reference frame, and if the observer moves inertially, we say that the observer has an inertial reference frame. A different way to state the first principle is to say that there is no such thing as a preferred inertial reference frame that is "at rest" in an absolute sense.

The second principle has to do with the way light moves. Light is an electromagnetic wave that consists of rippling electric and magnetic fields. When you turn on a source of light like a flashlight, the electromagnetic waves leave the emitting source and propagate away at an extraordinarily large but finite speed (about 300,000 km/s), much as if you tap the surface of a still pond, water waves propagate away from the point where you tapped.

Waves take on a life of their own after being emitted. By tapping the same point on the pond over and over again, you cannot influence the water waves that left earlier. In particular, the *speed* of the waves is not affected by the source. Whether you tap the pond repeatedly at the same point or you move your hand as you tap, the speed at which the resulting waves travel

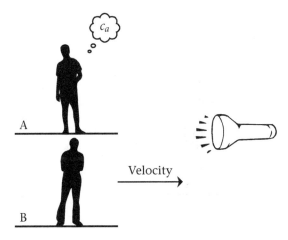

Figure 12.1
The situation from Person A's point of view.

is always the same. The crests of the waves might be closer or farther apart, depending on how you move your hand, but the crests always travel at the same speed, independent of the motion of their source. Since light is an electromagnetic wave, this observation also applies to light. In fact, that is precisely the second principle:

Principle II
The speed of light does not depend on the motion of its source.

These two principles may seem hardly surprising, maybe even mundane. But, taken together, they imply something radical.

Consider the following situation. Say there are two people, Person A and Person B, as well as a flashlight that has been switched on. Suppose that the flashlight is at rest with respect to A and that B is moving inertially toward the flashlight.

Person A can measure the speed of light in his reference frame (figure 12.1); we'll denote that measured speed as c_A. Person B can also measure the speed of light in her reference frame; we'll denote that measured speed as c_B. Person A isn't sure what B will measure. After all, B is moving against the propagating light. However, according to Principle I, Person B is equally entitled to think that she herself is at rest and that A and the flashlight are moving (figure 12.2). Person B will see a moving flashlight, which cannot affect the speed at which the light moves, according to Principle II.

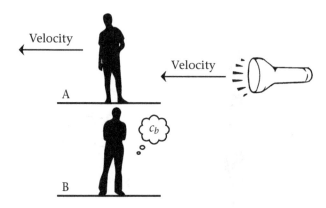

Figure 12.2
The situation from Person B's point of view.

Person B finds that the light moves at the same speed as reported by A. That is, $c_A = c_B$.

We can draw an even stronger conclusion. It was convenient for the argument, but Person B does not actually need to move inertially. If B moves noninertially, she may be able to tell that her reference frame is noninertial, and she will still see a moving source of light. A moving source of light still cannot affect the speed of light, even if its motion appears noninertial. Therefore, we arrive at an astonishing conclusion:

Principle I + II
The speed of light is the same in all reference frames.

According to special relativity, the speed of light is a fundamental constant, and we usually denote it by the letter c:

$$c = \text{the speed of light, about } 300,000 \text{ km/s}$$

You might be thinking: Why did it have to be light? Why doesn't the argument also apply to water waves? There must be something wrong, because water waves clearly appear to move faster or slower, depending on whether you travel against or with them!

The issue with water waves is that they do have a preferred reference frame: the reference frame in which the water appears stationary. Water waves cannot exist without the medium through which they travel—the water itself—which singles out a special reference frame that defines what it means to be at rest. The first principle, consequently, does not apply.

Light is special, in that it does not need a medium through which to travel, so there really is no preferred reference frame for the propagation of light. The fact that light needs no medium to propagate was tested experimentally by Albert Michelson and Edward Morley in 1887, even before Einstein proposed the theory of special relativity. In fact, the result of the Michelson-Morley experiment was one of the findings that guided Einstein toward relativity in the first place.

Before the era of relativity, physicists hypothesized the existence of a medium through which light propagated, which was referred to as the "luminiferous aether," or just "aether" for short (sometimes spelled *ether*). Because light travels through space, the aether would necessarily have to permeate the entire visible universe, and so the Earth constantly would be moving through the aether. Without going into too much detail, Michelson and Morley measured the time it took a beam of light to travel in various directions on Earth. Had there been an aether, they would have found small variations in the measured lengths of time, depending on how the Earth was moving with respect to the aether and whether the light beam was moving with or against the aether flow. To their surprise at the time, they found no difference in light's travel time, regardless of the direction. Michelson and Morley's experiment, the countless experiments that followed, and our modern theoretical understanding of physics all support the conclusion that there is no aether.

The principle that the speed of light is the same in all reference frames is what leads to the further conclusions that significantly differ from the intuition we develop in day-to-day experience. In particular, the phenomena of time dilation and length contraction are immediate consequences of Principles I + II, and knowledge of these phenomena is already enough to construct a plethora of paradoxes.

As the saying goes, a little knowledge is a dangerous thing.

12.1 Logical Implications

If the speed of light is the same in every reference frame, then we are forced to accept an unintuitive logical implication. To maintain that light always travels the same distance per unit of time in different reference frames, distances and durations must depend on the choice of reference frame itself. In fact, according to special relativity, we cannot really think of space and

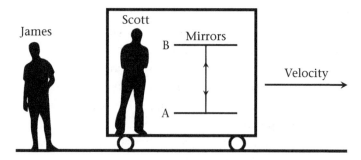

Figure 12.3
Scott and James's time dilation experiment.

time as independent concepts. They are two parts of a unified whole that we call *spacetime*. To a certain extent, you can think of the ambiguity in defining distances and durations as being due to the ambiguity in how to divide spacetime into space and time.

To begin, let's show that time *dilates* in a moving reference frame:

Time Dilation
A moving clock appears to run slower than a stationary clock.

Consider the following setup, as shown in figure 12.3. Suppose that Scott has two mirrors labeled A and B that are perfectly aligned and separated by a distance L. Suppose that a beam of light is bouncing between the two mirrors. Scott gets onto a movable platform in a laboratory with the mirrors, and then the platform begins to move inertially with speed v in a direction perpendicular to the axis of the two mirrors. Meanwhile, James stands still in the lab and watches Scott and the mirrors go by.

Next, Scott and James each measure the time it takes the light to make one round trip between the two mirrors (i.e., to go from Mirror A to Mirror B, then back to Mirror A again). From Scott's point of view (or in other words, in Scott's reference frame), the mirrors are stationary, so the light just goes up and down along a straight line, traveling a distance $2L$ at a constant speed c, as shown in figure 12.4.

Scott, therefore, thinks that the time taken for the light to make one round trip is

$$T_{\text{Scott}} = \frac{2L}{c} \qquad (12.1)$$

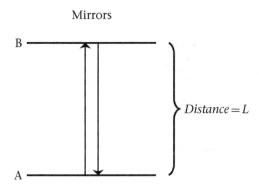

Figure 12.4
The light's trajectory in Scott's reference frame.

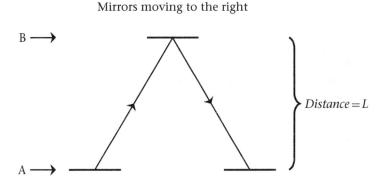

Figure 12.5
The light's trajectory in James's reference frame.

On the other hand, in James's reference frame, James sees the light move diagonally because from his point of view, the mirrors are in motion (figure 12.5).

James, therefore, sees the light travel a distance that is greater than $2L$ because the light also has to move horizontally. But, crucially, the light moves at the same constant speed, c. Because the light has to travel a longer distance at the same speed, James finds that the light's journey takes more time than what Scott finds. Using a bit of geometry, one can in fact show that

$$T_{James} = \gamma \, T_{Scott}, \qquad (12.2)$$

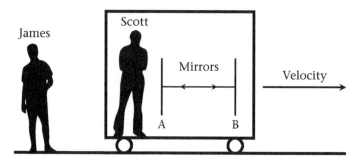

Figure 12.6
Scott and James's length contraction experiment.

where

$$\gamma = \frac{1}{\sqrt{1 - \frac{v^2}{c^2}}} . \tag{12.3}$$

(We define this function γ because it appears so often in special relativity; it saves us from having to write lots of square roots and fractions over and over again.) For example, if Scott is moving at 90 percent of the speed of light and James measures that a round trip takes $T_{\text{James}} = 1$ second, Scott finds that a round trip only takes $T_{\text{Scott}} = 0.45$ seconds. In other words, so far as James in concerned, Scott's clock is ticking slower than his. Time appears to run slower for Scott!

Another logical implication of Principles I + II is that the length of a moving object appears shorter relative to its length when it is stationary:

Length Contraction
A moving object's length along the direction of motion appears shorter than the length of the object when it is at rest.

Now suppose that Scott and James repeat their experiment, but this time, Scott aligns the mirrors along the direction in which the platform moves, as shown in figure 12.6.

In Scott's reference frame (figure 12.7), the situation is basically unchanged. The beam of light bounces between two stationary mirrors, so Scott still finds that the time for one round trip is

$$T_{\text{Scott}} = \frac{2L}{c} \tag{12.4}$$

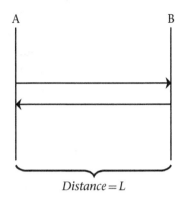

Figure 12.7
The light's trajectory in Scott's reference frame.

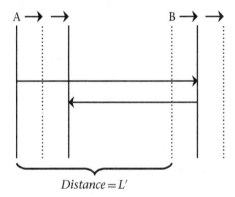

Figure 12.8
The light's new trajectory in James's reference frame.

The situation in James's reference frame (figure 12.8) is a bit more compli-cated. By the time the light reaches Mirror B, it will have moved a bit to the right, so on the trip from Mirror A to Mirror B, the light has to travel a bit farther than the length between the two mirrors. Then on the return trip, Mirror A moves toward the light, so the light has to cross a shorter distance than the length between the two mirrors.

If you carefully work out the geometry, you find that James measures the time for one round trip to be

$$T_{\text{James}} = \gamma^2 \frac{2L'}{c} \qquad (12.5)$$

where L' is the distance between the mirrors as measured by James. Notice that we did not write L in equation 12.5—instead, we used a different symbol, L'. This is because, according to what we know about time dilation, it must also be true that $T_{James} = \gamma \, T_{Scott}$, which, together with equations 12.4 and 12.5, means that

$$L' = \frac{1}{\gamma} L \tag{12.6}$$

In other words, it cannot be that James and Scott measure the same length between the two mirrors. James finds that the distance between the two mirrors is actually shorter than that when they are not moving at all.

An important detail to note is that length contraction happens only in the direction of motion. In the case where the two mirrors are aligned perpendicularly to the direction of motion, Scott and James find that the mirrors are separated by the same distance. In the perpendicular configuration, time dilation alone is enough to come up with a consistent explanation of the physics at play. In general, the operation that relates special relativistic quantities like times and positions in one inertial reference frame to those in another is called a *Lorentz transformation*, after the physicist Hendrik Lorentz, who was the first to figure out how these transformations work.

We tend not to have good intuition about special relativity because relativistic effects are not observable at conventional human speeds. For example, while moving at 100 km/h in a car, your clock loses only about 4×10^{-15} seconds, or 4 femtoseconds, for every second as measured by someone who watches you go by. In our day-to-day life, physics as described by Galileo and Newton tends to do a fine job of describing the world around us.

That said, we have reams of experimental evidence to back up the predictions of special relativity. In particle physics experiments, for instance, special relativity is indispensable. In particle accelerators, particles like protons and electrons are accelerated to such high speeds that relativistic effects are important. We directly see the effects of time dilation because unstable particles take longer to decay than they normally would when they are produced at high speeds in particle collisions. A similar thing happens for atmospheric muons. A *muon* is a type of unstable particle that is produced in large quantities in the upper atmosphere. Because it has a very high speed when produced, the average lifetime of an atmospheric muon appears longer to us than the lifetime of a muon at rest. This lets more muons reach us at the surface of the Earth before decaying than would otherwise be possible in the absence of time dilation.

At this point, we have all of the necessary groundwork to begin tackling paradoxes of special relativity. However, for the sake of completeness, it's worth closing this introduction by saying a few words about *general relativity*. Einstein's general theory of relativity builds on special relativity by incorporating gravity into the framework. The essential generalization is to upgrade the notion of inertial motion to *free-fall* motion. Without gravity, there is no difference between the two notions; however, in the presence of a massive body, only if you let yourself freely fall under the influence of gravity will you not feel any pushes or pulls as you move. With this definition, you can think of yourself as being *instantaneously* inertial at any given moment during free-fall.

In special relativity, inertial observers move along straight lines. In general relativity, since a freely falling observer is instantaneously inertial at any given moment, she can still think of herself as moving along straight lines. However, zooming out from a freely falling trajectory, the path might not look straight at all. For example, the planets trace out elliptical orbits as they freely fall around the Sun.

The only way to be moving in a straight line at any given moment while tracing out a trajectory that is not straight is if spacetime itself is curved. It's the same basic idea as traveling along a straight line on the Earth. If, at any given moment, you keep heading in the same direction, you will eventually come back to your starting point. You will do so having followed a trajectory that is definitely not straight, since the surface of the Earth is curved.

Einstein's extraordinary insight was that gravity *is* curvature. Massive bodies warp spacetime. Conversely, this curvature influences how objects move through space and time. It's a perfect cycle: Mass tells spacetime how to curve, and curvature tells mass how to move.

Nevertheless, we shall soon see that there's plenty to puzzle over, even without thinking about gravity.

12.2 Relativity of Simultaneity

An essential lesson from special relativity is that there is no absolute notion of a distance in space or an interval of time. Length and duration are relative concepts that make sense only after specifying a particular frame of

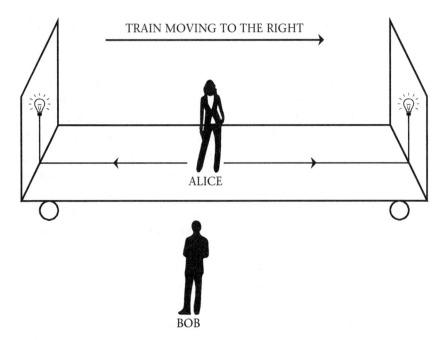

Figure 12.9
Bob watching Alice's train go by.

reference. Therefore, we should prepare ourselves for surprising conclusions when we scrutinize related notions, such as simultaneity, from a relativistic point of view. They are surprising, that is, until we have fully integrated and accepted the concepts of length and duration as relative.

The paradox illustrated in figure 12.9 is a variation on the classic question in special relativity of what it means for two events to be simultaneous. Einstein proposed a thought experiment along these lines during the early days of special relativity. Yet, despite its primordial origins, the relativity of simultaneity remains an important part of any complete discussion about special relativity.

Suppose that Alice is the conductor of a train that is moving at a constant speed along a perfectly horizontal and straight track. For a portion of her journey, Alice knows that the train will travel into a heavily forested area. So, at some particular moment, she presses a button located at the precise center of the train that sends a signal to each end of the train, turning on a light at the front and a light at the back. The signal is carried by an optical fiber, so assume that it travels at the speed of light. The lights, as well as the mechanisms connecting them to the button, are totally identical, so Alice sees the lights turn on simultaneously.

Meanwhile, Bob, who stands in a nearby park, watches Alice's train go by as the lights come on. Relative to Bob, Alice's train appears to be moving with a constant speed v from left to right. Let's also suppose that v is close to the speed of light, so relativistic effects are readily observable. The question, then, is what does Bob observe?

Claim 1

Bob also sees the lights come on at the same time. Alice and Bob could later meet in the next town and compare their observations. Therefore, they must see the same thing, lest they contradict each other regarding what actually happened.

Claim 2

Bob sees the taillight come on first, followed by the headlight shortly afterward. Because, according to Bob, the train is moving from left to right, the optical signal that turns on the taillight has to travel a shorter distance than the signal that turns on the headlight. Because the speed of light is the same in all inertial reference frames, it takes less time for the activation signal to reach the taillight than it does to reach the headlight.

Discussion and Resolution

Claim 2 is the correct resolution to this thought experiment. The conclusion that Alice and Bob will disagree on what they see may seem surprising at first. This is because, in our day-to-day experiences, we are accustomed to the idea that there is a global notion of time. We usually imagine that time marches forward for everyone, everywhere, and at the same rate. Simultaneity is an invariant notion according to this assumption. If everyone has the same clock, which displays the same reading, then it doesn't matter whose clock we use to decide the order of events. However, according to special relativity, every inertial observer has his or her own clock. When we realize that

- it is necessary to also specify which reference frame is used to determine the order of a set of events, and that
- different clocks in different reference frames tick at different rates, depending on their motion,

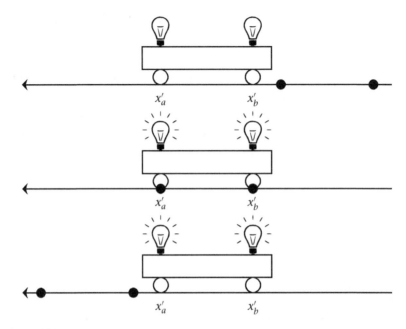

Figure 12.10
The train approaching the paint spots, from Alice's point of view.

it is less surprising that the order of events can depend on the reference frame.

While the short justification given in Claim 2 is correct, let's try to understand the physics even more concretely. To help us visualize what happens, suppose that there are also two spots of paint on the ground such that, in Alice's reference frame, the front and rear of the train are exactly over these points when the lights turn on. From Alice's point of view, she sees two spots of paint on the ground approach the train, and the moment at which they align with the front and rear of the train is when the headlight and taillight come on (figure 12.10).

Because, according to Alice, the ground below her is moving, the distance between these points is *contracted* relative to how far apart they are in Bob's frame of reference, in which the ground is not moving. Moreover, in Bob's reference frame, the length of Alice's train also appears contracted because her train is in motion. Now think about what Bob sees as he watches the train approach the paint spots. Naturally, the rear of the train will pass over the leftmost paint spot before the front of the train passes over the

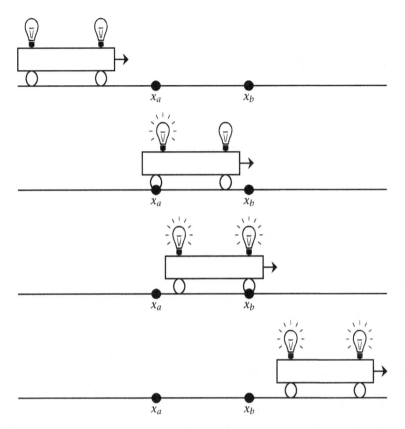

Figure 12.11
The train approaching the paint spots, from Bob's point of view.

rightmost spot, since the length of the train is shorter than the distance between the two paint spots in Bob's reference frame. Therefore, he sees the taillight turn on first, and then when the front of the train later makes it to the rightmost paint spot, he sees the headlight turn on (figure 12.11).

The technical reason why Alice and Bob can disagree on whether the lights are turning on simultaneously is that these two events are *spacelike-separated*. In other words, there is no causal path that connects the two events *a* and *b* in spacetime. When two events are spacelike-separated, the question of which event happened first is ambiguous. The answer depends on the reference frame. In fact, in the train example here, suppose there was another train on an immediately adjacent track moving from left to right (in Bob's reference frame) with a speed $V > v$. If Chan is aboard this train,

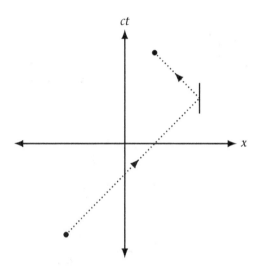

Figure 12.12
Two timelike-separated events (the two black dots) can be connected by a light ray and a mirror in spacetime.

then she sees Alice's train move in the opposite direction. The argument is consequently reflected, so she sees the headlight come on first.

A notion that *is* preserved in special relativity is causality. For example, while which light came on first on Alice's train depends on the reference frame, everyone in any frame will agree that Alice flipped the switch before the lights came on. Because the switch sends an optical pulse to turn on the two lights, the events "Alice flips the switch" and "The lights come on" are connected by light rays in spacetime. Such events are *lightlike-separated*. Events that can be connected by trajectories that move slower than the speed of light are *timelike-separated*.

Together, pairs of events that are lightlike-separated and timelike-separated are events that are causally connected. One event is always the cause and the other is always the effect, independent of the frame of reference. A way to understand this is by noting that any two causally connected events can be joined by future-directed light rays and mirrors, which are used to change the direction in which the light propagates. While different observers will report different distances and times between events connected by light rays, everyone will agree on the order of such events, provided that no one can exceed the speed of light. As such, the speed of light is a universal speed limit in special relativity. If it were possible to move faster than the speed of light, then causality would break down.

To help us visualize things like causality in special relativity, we often draw spacetime diagrams like the one shown in figure 12.12. In these diagrams, we explicitly draw the path that objects and light rays take in both space *and* time.

12.3 Twin Paradox

The Twin Paradox is another problem that challenges our intuition about time. Like the question of simultaneity, the Twin Paradox has existed in one form or another since Einstein's 1905 opus on special relativity.

Many centuries from now, John and Heather are twins who live on a space station for a long-term mission in deep space. One day, Heather boards a spacecraft and leaves to investigate a star system located a few light-years away. With her advanced propulsion technology, she travels to the star system at 90 percent of the speed of light, stops to collect some data, and then returns to the space station at 90 percent of the speed of light. Being twins, John and Heather were the same age before Heather left. However, by now, you are well aware that time is a delicate subject in special relativity, so let us ask how the twins' ages compare upon Heather's return:

- We know that time appears to move slower in moving reference frames. This is the principle of time dilation. So, from John's point of view, less time should appear to pass in Heather's reference frame.

- On the other hand, motion is relative. From Heather's point of view, it looks like John's space station receded and then came back toward her after she was finished collecting data. From her point of view, John's reference frame is the one moving, and the one in which time should move more slowly.

Even thinking relativistically, we are confronted with a paradox. John and Heather both start and finish at the same location, so how can they *each* think that the other sibling's clock was ticking slower?

Claim 1
John and Heather must still be the same age when Heather returns. This is the only possibility that is symmetric with regard to the relativity of motion. The answer does not depend on which sibling appears to be moving.

Claim 2

John is now older than Heather. Because John and Heather both start and finish at the same location in space but John remained stationary, this makes his reference frame the preferred one.

Discussion and Resolution

Claim 2 is the correct resolution of the paradox. The paradox is especially puzzling because the idea that motion is relative is a foundational aspect of this chapter. In this situation, however, the symmetry between who thinks who is moving is broken because Heather moves *noninertially*.

Recall that inertial reference frames are frames that are stationary or that move with a constant velocity. Any observer whose reference frame is inertial, such as John, is equally entitled to consider himself at rest. In contrast, Heather had to change both the direction and magnitude of her velocity in order to come to rest at the star system and subsequently return to the space station. In other words, at some point she had to fire the rockets on her spaceship (to slow down, to change direction, to gain speed), which exerted a *force* on her and caused her to accelerate. While there is no absolute notion of being at rest or moving with a constant velocity, the fact of whether you are accelerating *is* an absolute notion. Even if Heather could not see outside her spaceship, she would still be able to tell when she was accelerating when she felt the force being applied—just as you can feel a push when you accelerate in a car or take off in an airplane.

The resolution of the Twin Paradox is particularly clear if you draw John's and Heather's trajectories on a spacetime diagram (figure 12.13). Given any trajectory between two points in spacetime, imagine breaking it into a sequence of small translations in space, Δx, and small translations in time, Δt. When we work with time in addition to space, the right way to compute *spacetime* distance is to let each spatial translation Δx contribute a small amount of positive distance to the total (as usual), but to let each translation in time Δt contribute a small amount of negative distance. For example, John's trajectory does not move at all in space; he is always at the same location. Therefore, the length of his trajectory in spacetime is actually negative. This just means that John's trajectory was timelike, as discussed in the previous section. Moreover, the amount of time that elapsed for him is just the magnitude of this negative spacetime length.

By this logic, we see that even though Heather's trajectory through spacetime looks longer on paper, her trajectory has a shorter spacetime distance

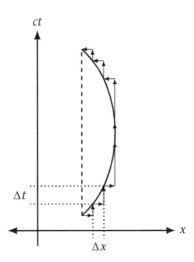

Figure 12.13
John's trajectory (the dashed line) and Heather's trajectory (the solid line) in spacetime.

than John's. She has to cover the same interval in the t-direction on the spacetime diagram as John does, but she also makes an excursion out in the x-direction. Her translations in space, Δx, counteract the translations that she undergoes in time, Δt, so her trajectory has a shorter duration in time.

An interesting variation to think about is the following. What if, instead, John remained stationary at a point in space and Heather moved at a constant speed around a circle over and over again, repeatedly meeting John at the top of the circle (figure 12.14). Heather is moving with respect to John, so we expect her time to be dilated, but she is moving at a constant speed. Does Heather age more slowly than John in this case as well?

The answer is yes! Even though Heather's speed stays the same, she has to accelerate to change the direction of her velocity in order to keep moving in a circle, and so her motion is noninertial. It turns out that there must always be a constant force acting on Heather and directed toward the center of the circle in order to keep her on this circular trajectory. For example, if you think about spinning a ball attached to a string in circles, then you apply a force on the ball through the string. It feels like you are always pulling the ball toward the center of its circular trajectory.

This example becomes interesting when we think about what Heather feels. If she stands on a platform as she goes around in circles, then she will

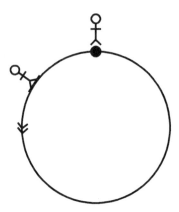

Figure 12.14
John is stationary while Heather moves at a constant speed in a circle.

feel the platform pushing back at her feet, similar to how you feel the ground pushing on your feet while standing on Earth. You may consequently wonder, as we stand here on Earth, do *we* experience time dilation relative to, say, someone on the International Space Station or out in space?

The answer is yes. If you were in a box with no windows and you felt a force pulling you toward the floor, could you distinguish whether you were on Earth or whether the box were accelerating upward? You could not. Because there is no distinction, and because we concluded that acceleration gives rise to time dilation, so must a gravitational field. It is well known in general relativity that observers in gravitational fields of different strengths experience different passages of time. Amazingly, this innocuous little thought experiment in special relativity has led us straight into general relativity.

12.4 Barn-Pole Paradox

Completing our trifecta of classic Einsteinian puzzles in relativity is the Barn-Pole Paradox.

Suppose that Rob is in possession of an indestructible pole 10 meters in length. He wants to store it in a barn on his property, but his barn is only 9 meters long. Fortunately, Rob is thinking like a relativist. The barn has two entrances, one at each end in the lengthwise direction. He aligns the length of the pole with the length of the barn, and he launches the pole toward the openings at $\sqrt{19}/10$ times the speed of light, which is precisely enough to cause the pole to contract to a length of 9 meters in his reference frame. The barn doors are equipped with indestructible shutters, which Rob closes at the exact moment when the pole fits entirely within the barn. Good for Rob. The pole is now in the barn!

Or is it? When the shutters close, the pole comes to rest. When the pole is at rest in Rob's reference frame, its length is 10 meters. It seems we are back to where we started. How can a 10-meter pole be trapped within a 9-meter barn? Did Rob actually succeed in closing the shutters?

Furthermore, if we consider the scenario from the pole's reference frame, it was the barn that contracted, not the pole. If the barn became even shorter relative to the pole, surely the pole could never have fit inside the barn at any point in time.

Claim 1

The pole's reference frame is the one that matters. In the pole's reference frame, the barn shrinks, so the pole will never fit inside.

Claim 2

There is an important distinction to be made between fitting the pole inside the barn and trapping the pole inside the barn. Whether the pole fits inside the barn depends on the reference frame. The pole can be trapped inside the barn, but not without shattering either the pole or the shutters immediately afterward; neither is actually indestructible, as the description suggests.

Discussion and Resolution

Claim 2 is correct. Because we do not deal with relativistic effects like length contraction and time dilation on a daily basis, we are naturally inclined to be suspicious of such effects when we encounter a paradox.

The Barn-Pole Paradox may seem familiar because it's actually a question of simultaneity. To begin, let's define what it means to fit and trap the pole inside the barn. Given a particular reference frame, we'll say that the pole fits inside the barn if both ends are simultaneously contained within the barn in that reference frame. Given some reference frame, we'll say that to trap the pole inside the barn is to have both barn shutters simultaneously held closed, while both ends of the pole are contained within the barn. We'll also name the two shutters: The "front shutter" is the one the pole reaches first, at the front of the barn; the "exit shutter" is the one at the back.

Fitting the Pole

The first aspect of the paradox deals with fitting the pole inside the barn. For this part of the paradox, we're not trying to keep the pole inside the barn, so the shutters are used only as benchmarks in time. Assume here that any time a shutter is closed, it is closed only for an instant before immediately reopening.

Does the pole ever fit inside the barn? The answer depends on the reference frame. In the barn's frame, the answer is yes. The pole contracts so that it fits in the barn, and the shutters close simultaneously—as viewed by the barn—the moment the pole is contained entirely between the shutters. Then the shutters open again, simultaneously, as viewed by the barn.

But, as we know from our study of the relativity of simultaneity, the shutters will not close and reopen simultaneously in the pole's reference frame. In this frame, the pole is in fact much longer than the barn. We can map the situation directly onto Alice and Rob's signaling experiment from section 12.2. The shutters' closing is just like the lights going off on the train if we think of the pole as the reference frame that is "at rest"—which we are entitled to do when thinking relativistically. In the pole's reference frame, the exit shutter will close, then immediately reopen, first when the front of the pole reaches it. The front shutter closes and immediately reopens later, when the end of the pole catches up to the front shutter, by which time the front of the pole has sailed well beyond the exit.

Trapping the Pole

The next question is whether the pole can ever be trapped inside the barn.

In the barn's reference frame, the answer is a straightforward yes. In this frame, the pole contracts so as to fit within the barn. The moment it fits, Rob closes both shutters, and the pole instantaneously stops when the front of the pole hits the exit shutter. Now that the pole is no longer moving relative to the barn, it is no longer contracted. It returns to its original size of 10 meters in a 9-meter barn, and in so doing, it shatters. The assumption that both the pole and barn shutters were indestructible, as given in the setup, was a false one.

Now, this story gives rise to an altogether new paradox. The simultaneous closing of shutters in the barn's reference frame caused the pole to shatter.

Rob was only able to shatter the pole by trapping it inside the barn. If the pole shatters in one frame, it must shatter in all frames—as a result of getting trapped inside the barn—but in the previous section, we learned that in the pole's reference frame, the pole never fits inside the barn. In the pole's frame, how can the pole get trapped if it never fits?

Let's consider the entire scenario from the pole's frame. The first thing to happen is for the front of the pole to reach the back of the barn, at which time the exit shutter closes (now permanently, without reopening). In the barn's frame, every fiber of the pole stops instantaneously and simultaneously upon contact with the shutter, but in the pole's frame, the pole's deceleration is not simultaneous. The front of the pole decelerates first, and the deceleration ripples from the front of the pole all the way to the back. By the time the back of the pole decelerates, it has caught up to the barn's front door, at which point the front shutter closes and the pole shatters.

All in all, Rob would do better by finding a slightly bigger barn or clipping off the end of his pole.

12.5 Dewan and Beran's Paradox (Bell's Spaceship)

Dewan and Beran's Paradox was initially proposed and resolved by Edmond Dewan and Michael Beran. However, John Stewart Bell's discussion of the problem seems to have garnered a higher readership over time, so this paradox is more popularly known as "Bell's Spaceship."

Suppose that Ann is an inertial observer; her frame is what we will call "at rest" in this problem. She has prepared two spaceships that are pointing in the same direction and are connected by a taut string. The spaceships are initially at rest, then at a particular moment, they both fire their rockets and begin accelerating at the same rate so that, according to Ann, they have the same speed at any given later time (figure 12.15). By this logic, the distance between the spaceships should stay the same in Ann's frame, as should the length of the string.

However, recall that moving objects exhibit length contraction. According to this latter observation, the string should appear length-contracted to Ann, and it should keep getting shorter and shorter as the spaceships move faster and faster. We seem to have a paradox. What happens to the string and the distance between the spaceships in Ann's reference frame?

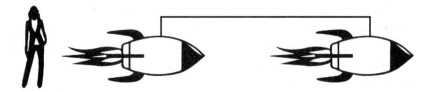

Figure 12.15
Ann, and two ships accelerating from rest at the same rate.

Claim 1

The string is in motion, so it must appear length-contracted to Ann. The distance between the spaceships therefore also appears shortened according to Ann.

Claim 2

The distance between the spaceships remains the same according to Ann. Because the spaceships and the string are accelerating, or more pertinently, not moving inertially, the usual rules of length contraction do not apply.

Discussion and Resolution

Neither claim is quite right, but Claim 2 is closer to being correct. Rather, Claim 2 is technically correct but lacking in details. It is true that a reference frame attached to either accelerating spaceship is not inertial. Nevertheless, consider the following small modification to the problem. We could suppose that the rockets switch off after a fixed amount of time has passed in Ann's reference frame. The spaceships will always be the same distance apart in Ann's frame, but after this point, the rockets *are* moving inertially. Therefore, we really need to understand how length contraction is consistent with this experiment. (The point is that acceleration is not a miraculous panacea in special relativity—some thought is still required.) The string is also somewhat of a distraction in this problem. It causes us to think about messy material properties like stretching and tension, which only mar our understanding of the problem. Therefore, let's forget about the string for now, and idealize the spaceships down to a pair of moving points.

Without the string, the accelerated trajectories through spacetime of the two points representing the spaceships could look something like those

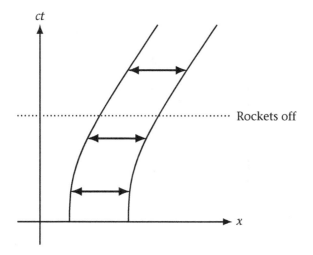

Figure 12.16
The trajectories of the two spaceships, idealized as points, as viewed in Ann's reference frame. Notice that the two ships always have the same horizontal separation in this frame.

shown in figure 12.16. The coordinate system in the diagram is made up of Ann's coordinates. Indeed, at any fixed time according to Ann, the corresponding points on each trajectory have the same horizontal separation, and so the spaceships always appear to be separated by the same distance according to Ann.

But what happens according to the spaceships? Consider first the tailing spaceship. Because it is accelerating, it does not have a single inertial frame attached to it throughout its motion. However, at any given moment, we can consider an instantaneously inertial frame attached to it. That is, at a given moment when the spaceship is moving with speed v relative to Ann, attach to the ship an inertial frame that moves with the same speed v in the same direction only for that moment. In this instantaneously inertial frame, we can ask what events are instantaneously simultaneous for the tailing spaceship. These instantaneously simultaneous events form a straight but slightly sloped line (whose slope is a function of v). In particular, notice that this line intersects the trajectory of the leading spaceship farther along its trajectory than the point corresponding to where the tailing spaceship is on its trajectory. In other words, to the tailing ship, the leading ship appears farther away than when they started accelerating from rest! If the

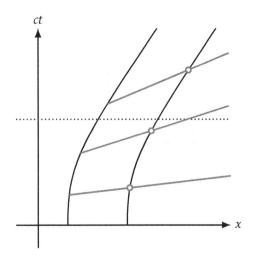

Figure 12.17
The gray lines are instantaneous lines of simultaneity for the tailing spaceship. Each line intersects the leading ship's trajectory farther from the corresponding point where the line of simultaneity originated on the tailing ship's trajectory.

spaceships continue to accelerate, then at a later point on the tailing ship's trajectory, we can draw another instantaneously inertial frame. The ship will be moving faster, and so its line of instantaneously simultaneous events will be even more sloped, meaning that the leading ship will appear to be even farther away (figure 12.17).

It should be clear what's happening now: While the spaceships are accelerating, from the point of view of the tailing ship, it looks like the leading spaceship is pulling ahead. Conversely, from the point of view of the leading ship, it looks like the tailing ship is falling behind. Put another way, the events "the tailing spaceship has speed v" and "the leading spaceship has the same speed v" are simultaneous in Ann's reference frame. Therefore, these events cannot be simultaneous according to either of the spaceships. It follows that at any given time on one spaceship's clock, that spaceship will see the other moving at a different speed.

Now we can also make sense of how length contraction works in this problem. If the ships stop accelerating, then they will be moving with a constant velocity, so they will indeed subsequently coexist in a single inertial reference frame. However, because the leading ship was pulling ahead according to the tailing ship (or equivalently, because the tailing ship was lagging behind according to the leading ship), in the postacceleration inertial frame, the ships will be farther apart than when they began accelerating. It turns out that this increase in distance, as measured in the new moving inertial frame, is precisely undone by the length contraction that is brought on by going back to Ann's stationary frame. So length contraction still happens—it's just that the spaceships got farther apart in the moving frame, so that the contracted distance separating them looks unchanged to Ann!

In closing, let's come back to the question of what happens to the string. Here, the answer is really: "How realistically do you want to model the string and its material properties?" Michael Weiss and Don Koks published an excellent discussion of the various subtleties that arise when modeling the string with increasing physical fidelity. Nevertheless, at a coarse level, because the spaceships get farther apart as measured in a collection of instantaneously inertial frames, it's probably safe to say that the string will stretch and eventually snap if the spaceships accelerate for too long.

12.6 Ehrenfest's Paradox

In a short article penned in 1909, Paul Ehrenfest proposed the following paradox that concerns rotation and rigidity in special relativity. By now, the paradox is well understood; nevertheless, people continue to discuss it and reformulate new resolutions to this day. The fact that Ehrenfest's Paradox endures as a fruitful thought experiment highlights the complexity of rotational motion in relativity and the subtle distinctions that must be made in the discussion of this topic, such as between the motion of a rigid object and the rigid motion of an object.

Suppose that Loie has a solid disk of radius R (say $R = 1$ meter, but the precise length is not important). She then slowly starts to spin up the disk, taking great care to apply forces in just the right way so as not to deform or warp it in any way. After a while, suppose that the disk reaches a steady rotational frequency such that points on the edge of the disk are moving with a speed that is close to the speed of light.

What happens to the dimensions of the disk? First, consider the radius. In Loie's frame of reference, has it changed to some new value R_{new}? The answer is no: While indeed every point on the disk (except for the exact center) is now moving with some nonzero speed, a straight radial line drawn

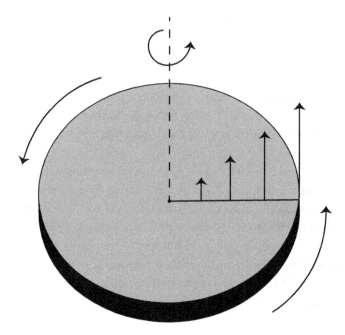

Figure 12.18
Loie's spinning disk. Notice how at any given moment, points on a line drawn from the center of the disk to its edge have velocities that are perpendicular to that line.

from the center of the disk to the edge is everywhere perpendicular to the direction of motion (see figure 12.18). There is no length contraction in the radial direction, so $R_{\text{new}} = R$.

However, consider the disk's circumference. Because the rim of the disk is moving with a very high speed and a line drawn along the rim of the disk *does* coincide with the direction of motion, we should expect that the circumference is now shorter due to length contraction. In other words, the new circumference, $2\pi R_{\text{new}}$, should be less than $2\pi R$, or dividing by 2π, R_{new} should be less than R. It would seem that we have a paradox—it can't be that both $R_{\text{new}} = R$ and $R_{\text{new}} < R$!

Claim 1

R_{new} is the same as R, but the circumference is actually shorter. Who said that geometry had to be Euclidean? Spinning up the disk warped the underlying geometry of space itself so that circumference is no longer $2\pi r$.

Claim 2

No, geometry really must be Euclidean because gravity is not a part of this thought experiment, so $R_{new} < R$. We are doing special relativity, after all! It must not be possible to spin the disk without changing its radius.

Discussion and Resolution

Claim 2 is indeed correct (although spinning objects are also important for general relativity when gravity is concerned). The logical inconsistency that sneaked in is the idea of the solid and undeformable, or *rigid*, disk.

An ideal rigid object cannot be compressed, stretched, or otherwise deformed. It is unmalleable and inflexible, and it perfecly transmits shocks and forces throughout its extent. Many objects in our day-to-day experiences are, to a very good approximation, rigid. Just imagine how hard it is to deform a tempered steel beam! But while you could argue that there is no such thing as a truly rigid object in even a realistic Newtonian description of the world, at least rigid objects are not logically inconsistent with the principles of Newtonian mechanics. According to Newtonian mechanics, it is perfectly fine to imagine a rigid object, which you can move in any way and subject to whatever forces and accelerations you like. It turns out that this idealization breaks down in special relativity.

As alluded to at the beginning of this section, an important distinction must be made between rigid objects and rigid motion, because while the idea of a truly rigid object does not exist in special relativity, it is possible for an object to *move* rigidly.

Rigid motion is roughly defined as follows. Given some object that is in motion, pick any point on the object, label it P, and follow the trajectory that it traces out through spacetime. If the distance between this trajectory and that traced out by any other nearby point remains the same, then the object is said to be moving rigidly. This definition lacks some technical details, but the gist is correct, and it coincides with our Newtonian intuition.

An important question we cannot overlook, however, is: "What reference frame is used to compute the distance between the two trajectories?" The answer is: "An inertial reference frame that moves with the same instantaneous velocity at a given moment on the point's trajectory." That is, at any moment on P's trajectory, the point P will have some velocity. So, use an inertial reference frame that is centered on P and that is moving with exactly the same velocity, but just at that moment. At any other moment,

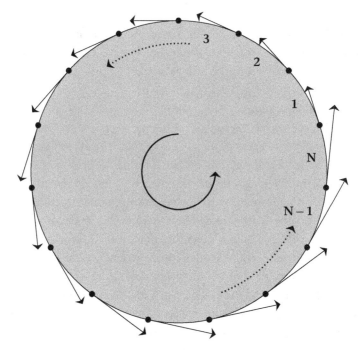

Figure 12.19
Divide Loie's disk into N segments and imagine speeding it up a bit. The acceleration of the leading end point of any segment must be larger than the acceleration of the segment's tailing end point.

you would need to use a different reference frame with the corresponding velocity, but that's the prescription. This sort of reference frame is called an *instantaneously inertial frame*.

It turns out that it's perfectly fine for a disk to rotate rigidly, but crucially, it's impossible to spin up a disk from rest in a rigid way. This is what we mean when we say that there is no such thing as a rigid disk. There exists no disk, in theory or practice, that can be spun up such that its motion is rigid at all times. We can see this through the following argument, which is a variation on a resolution of the paradox by Grøn.

Consider Loie's disk, and imagine dividing the edge of the disk into many small segments, as shown in figure 12.19. (Technically, we need to consider a limit where the number of segments tends to infinity.) The disk can be at rest, or it already can be rotating. Either way, suppose that we want to spin the disk up a little bit more, but *rigidly*. This means that, for any of the segments, the length of the segment cannot be changing in its instantaneously inertial reference frame. This in turn means that the end points of any given segment must be accelerating in the same way in their instantaneously inertial reference frame.

This should sound quite familiar: In their instantaneously inertial frames, each segment is like a miniature version of Bell's spaceship! Now let's ask what this implies in Loie's reference frame. Suppose that we number the segments from 1 to N, where N stands for the number of the last segment, and the numbering is going in the direction in which we want to spin up the disk. Using what we know about Bell's spaceship, if the end points of the first segment have the same acceleration in their instantaneously inertial frame, then in Loie's frame, the leading end point must have a larger acceleration than the tailing end point. However, the leading end point of segment 1 is the tailing end point of segment 2, so the leading end point of segment 2 must have a larger acceleration than the very first tailing end point. But, the leading end point of segment 2 is the tailing end point of segment 3, and so on. Once we reach the last end point, we arrive at a contradiction: The leading end point of segment N must have a larger acceleration than the very first tailing end point, but these are one and the same point. We have made it all the way back to the start of the circle! We see, therefore, that it is *kinematically* impossible to spin up the disk in a rigid way.

And so Ehrenfest's Paradox is resolved. While Loie could very well come into possession of a disk that is rigidly rotating, it could never have gotten to that state by starting from rest in Loie's reference frame and remaining rigid at all times. At some point it gets deformed radially.

12.7 Supplee's Paradox

Supplee's Paradox was proposed by James M. Supplee in 1989. It's different from the preceding paradoxes, in that it deals directly with the forces that influence an object's motion instead of just the motion itself. *Forces* are the pushes and pulls that cause objects to accelerate and deviate from inertial trajectories, and we will find that we must think carefully about how forces behave in a relativistic setting.

In particular, Supplee's Paradox has to do with the buoyant force that an object experiences that makes it float when it is submersed in a fluid. A *buoyant force* arises when there is a difference in pressure between the

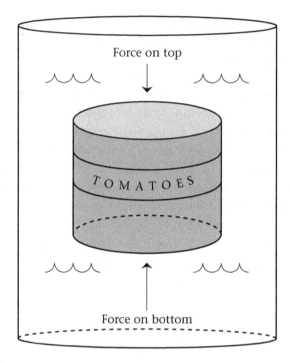

Figure 12.20
Higher water pressure at the bottom of the tin produces a bigger force there than at the top of the tin.

top of the object and its bottom. For example, if you hold a sealed tin of tomato sauce upright in a container full of water, the pressure is higher at the bottom of the tin than at its top due to the weight of the water in between. Therefore, the force due to the pressure of water pushing up on the bottom of the tin is larger than the force pushing down on the top of the tin, so there is a net buoyant force directed upward (figure 12.20). (The forces on the sides of the tin cancel out—for any point on the side where the water exerts a force, the same force is exerted in the opposite direction at the diametrically opposed point.) If the tin is full, then of course it won't float, but it will feel a bit lighter than when you hold it in air. This is also why your body feels lighter when you stand in a pool of water.

As for how big the buoyant force actually is, we can thank the Greek philosopher-mathematician Archimedes for uncovering that piece of knowledge in the third century B.C. Archimedes' principle famously states that the buoyant force on an object is equal to the weight of the water displaced

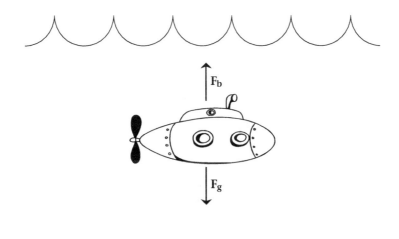

Figure 12.21
A submarine at rest. Its density is the same as that of water, so the buoyant force, F_b, exactly counteracts the force due to gravity, F_g.

by the object—or in other words, the weight of the water that would have been there if the object were not. The buoyant force is bigger in denser fluids. That's why it's easier to float in salt water than in fresh water, which is less dense.

Supplee's Paradox is as follows: Consider a submarine at rest in a body of water on Earth. Suppose that the density of the submarine has been tuned to match the density of water so the weight of the submarine is exactly equal to the weight of the water it displaces. Archimedes' principle then implies that the buoyant force, which makes the submarine float, perfectly counters the downward pull of gravity on the submarine (figure 12.21). The submarine just sits in place, neither sinking nor rising. Here, we will neglect the curvature of the Earth so that we can model the gravitational field as being uniform and directed downward. We will also ignore hydrodynamic effects like viscosity or turbulence.

Now suppose that the submarine starts moving to the right and eventually reaches a large-enough constant speed that relativistic effects are important. What happens to the buoyant force acting on the submarine? Because it is now moving, the submarine's length appears contracted. This means that it now occupies less volume than it did before, and so it's displacing less water than it did before. It must be that the buoyant force acting

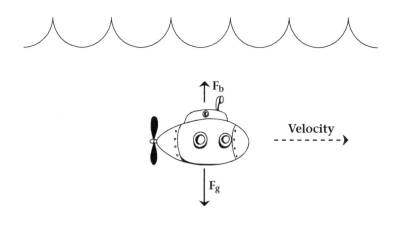

Figure 12.22
The submarine starts to move to the right. The buoyant force is now smaller.

on the submarine is consequently smaller, and so we conclude that the submarine should start to sink (figure 12.22).

However, what if instead, we consider an inertial frame moving with the submarine, in which the submarine appears stationary and the water appears to be flowing to the left? In this frame, the length of the submarine is unchanged, but because the water is moving, its density is higher due to length contraction. In this frame, 1 gram of water takes up a bit less space than 1 cm^3. The weight of the water displaced by the submarine is consequently larger, and so according to Archimedes' principle, the buoyant force should be greater as well (figure 12.23). We therefore conclude that the submarine should start to float.

It most certainly cannot be the case that the submarine floats in one frame of reference and sinks in the other, so we have a paradox!

Claim 1
The submarine floats, so the buoyant force must have a relativistic correction due to the submarine's motion in the stationary frame.

Claim 2
The submarine sinks, so the buoyant force must have a relativistic correction due to the motion of the water in the moving frame.

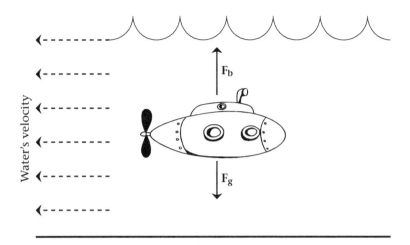

Figure 12.23
From the point of view of the submarine, the water is flowing to the left. In this frame, the buoyant force is apparently greater.

Discussion and Resolution

The key to resolving this paradox is understanding that forces transform in a complicated and somewhat unintuitive way when going to a different inertial frame. Roughly, this is because a force acting on an object causes its velocity to change over time, and both velocity and time depend on the choice of reference frame. A careful mathematical analysis would reveal that Claim 2 is correct. In fact, it turns out that Archimedes' principle isn't quite right in the moving frame; the buoyant force is not just equal to the weight of the displaced water. Essentially, the moving water pushes a little bit less on the submarine, in a way that counteracts the density increase so that the submarine sinks. A precise analysis of why this happens is unavoidably mathematical; however, we can reason why the buoyant force should change using some principles of relativity that we already know, as well as why the change should be a reduction in the buoyant force's size compared to a naïve application of Archimedes' principle.

Recall that the buoyant force is due to a pressure difference. Because of this fact, it turns out that the quantity that is important for computing the buoyant force is the rate of change of pressure as a function of position and depth, or in other words, the *pressure gradient*. In fact, you can use some tools from calculus to show that the pressure gradient is directly responsible

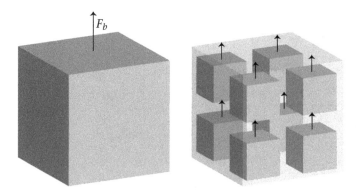

Figure 12.24
Using calculus, you can think of the total buoyant force as being the sum of contributions from a bunch of little packets of displaced water—the little cubes on the right.

for Archimedes' principle. If you divide up the fictitious displaced water into a number of small packets and add up the forces acting on each packet due to the pressure gradient, you obtain the buoyant force and can show that Archimedes' principle follows, at least nonrelativistically (figure 12.24).

Notice that pressure itself has no intrinsic direction attached to it. It's just a number, so it's not a geometric quantity that changes depending on the frame of reference. Also note that, in this setting, the water pressure varies only as a function of depth, and depth is not contracted because it is perpendicular to the direction of motion. Therefore, it follows that the pressure gradient is the same in both the stationary frame and the frame that moves with the submarine. The ramification of this observation is that a packet of moving water in the frame that moves with the submarine contributes the same amount to buoyancy as a packet of stationary water in the stationary water, even though the moving water is denser. We therefore conclude that the buoyant force must receive a correction in the moving frame, which makes it smaller than what one would predict using Archimedes' principle.

Outside of Supplee's Paradox, understanding buoyancy relativistically has important applications in stellar hydrodynamics. Stars are not made of water, but we can still investigate how their constituent material flows. It's also possible to reformulate Archimedes' principle to get a relativistic version that is still valid in special relativity, where the weight of the displaced volume of water (but not its mass) depends on the speed at which it flows.

13 Quantum Mechanics

Michael Coughlin, Matt Cook, and Aidan Chatwin-Davies

> I cannot define the real problem, therefore I suspect there's no real problem, but I'm not sure there's no real problem.
> —Richard Feynman

The early twentieth century saw the rise of what would become two of the most important physical theories: general relativity and quantum mechanics. In chapter 12, we touched on Albert Einstein's theory of general relativity only very briefly; however, we should not let that discount the importance of this theory for our understanding of how the universe works. Einstein's insight that gravity is curvature—that massive bodies literally twist, pull, and stretch space and time, and in turn they move in response to these deformations—redefined the way we think about the fabric of the universe. Moreover, just as special relativity has passed many experimental checks, so too has general relativity.

The first experimental success of the theory was its correct prediction of how Mercury's orbit around the Sun changes in time, which is a phenomenon that had been observed over centuries but had eluded explanation (the so-called precession of the perihelion of Mercury). Another more stringent test was the prediction that the Sun's gravity should deflect the light of stars that are behind the Sun from our point of view here on Earth; this is called *gravitational lensing*, and it was first measured by Sir Arthur Eddington during an eclipse.

Many other astrophysical phenomena have provided us with experimental evidence that is consistent with general relativity, including, most recently, the direct detection of gravitational waves by the Laser Interferometer Gravitational-Wave Observatory (LIGO). Gravitational waves are ripples in spacetime that are produced when two massive bodies that were initially orbiting each other, such as pairs of heavy stars or pairs of black holes, spiral into each other and collide. Violent collisions of massive objects can cause spacetime ripples, which eventually reach us here on Earth.

Then, there's quantum mechanics. In some ways, that theory is quite similar to general relativity. It changed the way we think about the universe, succeeded in explaining many previously mysterious phenomena, and has withstood excruciatingly precise experimental scrutiny. In fact, in terms of its predictive power, quantum mechanics is one of the most successful theories of physics. However, in other ways, the theory couldn't be more different from general relativity.

Perhaps the most blatant difference is that quantum mechanics is a theory of the very small. While general relativity deals with space, time, and the universe on large scales, quantum mechanics is the language that we use to describe the physics of submicroscopic objects like atoms, electrons, quarks, and smaller things that we may not even know about yet. Another difference is a sociological one. While the foundations of general relativity were largely laid out by Einstein, quantum mechanics was developed by many people, with many stops and starts along the way.

The story of quantum mechanics begins at the turn of the twentieth century with Max Planck, who was a professor of theoretical physics at Berlin University who would later go on to win the Nobel Prize in Physics for his work. Planck's goal was to understand the amount of light emitted by an object at a particular temperature. All objects with a nonzero temperature emit heat in the form of light. That's why a stove element glows red when it heats up. Even at room temperature, everything is constantly emitting light, although we don't see it because most of this light is emitted in the infrared range. (Some types of night vision goggles help us see this light, which is the basis for how they work.) At even lower temperatures, objects emit mostly at even lower energies; for example, an object floating in the coldness of space emits mostly microwaves.

We might not usually think of microwaves as light, but microwaves and light, as well as radio waves, x-rays, and gamma rays, are really all just *electromagnetic waves*—oscillating electric and magnetic fields. The only difference is the lengths between the crests of the waves, which we call *wavelengths*. Long wavelengths correspond to low-energy electromagnetic waves, while short wavelengths correspond to high-energy electromagnetic waves. For example, a typical wavelength for microwaves is 1 centimeter, while a typical wavelength for red light is 675 nanometers. The wavelength of infrared light is a bit bigger than that of red light, hence the term *infrared*.

Planck was interested in characterizing how a black body emits light. Physicists use the term *black body* to describe an idealized, perfect object that absorbs all the electromagnetic radiation that hits it. The classical understanding was that an ideal black body at a constant temperature would smoothly absorb and radiate energy across all wavelengths of electromagnetic radiation. The problem was that, for technical reasons, this smoothness implied that the amount of energy that the black body emits in radiation of a given wavelength would get bigger and bigger for shorter and shorter wavelengths. This would mean that black bodies emit an infinite amount of energy, which is abhorrent on theoretical grounds and certainly not observed experimentally. This failing of classical physics is known as the *ultraviolet catastrophe.*

Planck found that if, instead of being able to absorb or emit any amount of energy, a black body could only absorb or emit energy in discrete chunks that he called *quanta*, then the ultraviolet catastrophe disappeared. Correspondingly, any such quantity that only comes in chunks like this is said to be *quantized*. Planck's quantization scheme changed the black-body spectrum significantly, causing the amount of energy emitted at a given wavelength to approach zero as the wavelength approached zero. This fixed the ultraviolet catastrophe, but while successful at explaining experimental observation, the reasons for its success were not understood.

Planck's insight was useful for solving another mystery of classical physics: the stability of atoms. Consider, for instance, the hydrogen atom. Hydrogen is the simplest of atoms, consisting of a single positively charged proton that is orbited by a (much smaller) negatively charged electron. Classically, this setup leads to a problematic conclusion. The theory of electromagnetism predicts that an electrically charged object moving in a circular path through an electric field will radiate energy in the form of light. In the case of the hydrogen atom, the electron is a charged object that moves in circles through the electric field produced by the proton. In the classical view, therefore, the electron should radiate as it orbits the proton; however, as the electron gives up energy to produce radiation, it should spiral into the proton in a matter of picoseconds, with the two of them combining to produce a neutron when they meet.

Clearly, this inward spiral doesn't happen because the atoms around us—and of our own bodies—aren't all disintegrating. Inspired by the ideas of quantized energy, Niels Bohr, a lecturer at the University of Copenhagen (who would also later win a Nobel Prize in Physics), proposed that electrons are quantized in such a way that they can orbit only at a discrete collection of specific radii around the proton. An electron could then jump from orbit to orbit by emitting quanta of energy in the form of electromagnetic radiation. In particular, Bohr postulated that an electron simply couldn't emit the quantum of energy that it would have to do in order to fall into the proton. At the time, he wasn't entirely sure why this would be the case, but the idea of quantized orbits in which jumping to the proton is forbidden certainly stabilized his model of the hydrogen atom.

Einstein synthesized the successes of Planck's and Bohr's quantum theories by proposing that it was not just the light emitted by individual atoms or black bodies that was quantized, but in fact *all* electromagnetic radiation. This was a radical proposal; at the time, it was firmly believed that electromagnetic radiation was wave-like. Einstein was proposing that light, and all electromagnetic radiation, has intrinsically particle-like properties. Quanta of light are what we now call *photons*. Einstein was later awarded a Nobel Prize in Physics for his work on the physics of photons.

If light can behave like a particle, can particles of matter behave like waves? This is precisely the question that was pondered by Louis de Broglie, a PhD student at Paris-Sorbonne University and later a Nobel laureate as well. According to Einstein's theory of special relativity, a photon has momentum, p, given by

$$p = \frac{h}{\lambda},$$

where h is a constant number known as *Planck's constant* and λ is the photon's wavelength. Nothing stops us from rewriting this equation to solve for the wavelength:

$$\lambda = \frac{h}{p}$$

Here, de Broglie's bold suggestion was that this equation should make sense even for massive particles, whose momentum is given by their mass times their velocity. Moreover, he suggested that there were circumstances in which a particle would exhibit wave-like properties, as if it were a wave with a wavelength equal to h/p.

The phenomenon of waves behaving like particles and particles behaving like waves is what we call *wave-particle duality*. It's one of the surprising things that quantum mechanics has taught us about the submicroscopic world: It seems that the submicroscopic constituents of the universe are neither absolutely particles nor absolutely waves. They exhibit characteristics of both. Some of the consequences of wave-particle duality will be the subject of the first paradox that we encounter in this chapter.

Wave-particle duality is one of the historical reasons why we call the thing that describes the quantum-mechanical state of an object a *wave function*. Rather than going into precise details about wave functions, consider, for example, a particle like an electron. Classically, if you wanted to describe its position, then all you would have to do is provide a single location—a set of three coordinates when you have three dimensions of space. In quantum mechanics, the electron has a wave function, which you can think of as a profile over *all* possible locations where it could be. This thing behaves a lot like a wave and characterizes many of the wave-like properties of objects that we would otherwise call *particles*.

Mathematically, how we calculate and manipulate wave functions is unambiguous, but how the mathematics correspond to reality remains a source of debate today. Physicists have different interpretations of the wave function, and thus of quantum mechanics as a whole. One of the original interpretations was made by the German physicist and Nobel laureate Max Born, who viewed the wave function as describing the probability of measuring a particle in a particular portion of space. So, in our example, if the wave function of our electron has peaked around a particular location in space and is almost zero everywhere else, then with very high probability, we can say that the electron is located at that position, and it approximately behaves like a particle. Other interpretations will be described in Paradox 13.2, described in section 13.2 on Schrödinger's Cat.

On the other hand, what if the wave function of a particle is evenly spread over a large region? Then, according to Born's interpretation, we can't really say where the "particle" is with any confidence at all. *Uncertainty* is another strange, and perhaps even disturbing, feature of quantum mechanics. The theory is inherently probabilistic. To make things even more perplexing, many quantum quantities come in pairs, such that the more that is known about one, the less is known about the other. For one such pair of quantities (namely, position and momentum), this is formalized by the *uncertainty principle* of Werner Heisenberg (physicist and Nobel laureate). His principle states that the more information you have about a particle's position, then the less precisely you can characterize its momentum, and vice versa.

Given a particle that has some wave function, however, if you were to measure its position, then you eliminate the uncertainty in its location. This shouldn't be so surprising; you checked where the particle is located, so now you know where it's located. The surprising observation is this: Because the information about the particle's location has been modified, your measurement must have somehow influenced the object that contains this information—namely, the particle's wave function. In other words, the act of measuring a quantum system influences the system itself. Measurement and observer dependence will be the subjects of several of the paradoxes in this chapter.

To recap, we have seen that the wave function, uncertainty, and observer dependence are three surprising features of quantum mechanics. The three of these conspire to make possible a phenomenon that is possibly the most quantum feature of quantum mechanics: entanglement. Essentially, the properties of quantum mechanics can be leveraged to create correlations between quantum systems that are stronger than any classical correlations. These correlations are so strong that, in theory, measuring one part of an entangled system influences the other part, even if it's on the other side of the visible universe—and this happens seemingly instantaneously.

Entanglement was particularly vexing to Einstein. Instantaneous "action at a distance" is not something that should be possible according to relativity. This led him to pose a paradox that we will take up in section 13.4—and one that we will see resolved. In the end, entanglement can be reconciled with relativity, although perhaps not as nicely as Einstein would have hoped.

You are now prepared for the four paradoxes that will be discussed in this chapter. But first, let us return to the tension between general relativity and quantum mechanics. These two disciplines have profoundly affected our understanding of physics, both for theoretical predictions and experimental checks. A major open problem in physics, however, is unifying these two theories into a single theory of quantum gravity. This is a difficult task for deep mathematical and conceptual reasons.

The essence of the challenge is that quantum mechanics usually relies on the existence of space and time—but in quantum gravity, spacetime itself has to be a quantum object. Is it possible to write a consistent wave function for space and time? What does uncertainty mean for geometry? What role does entanglement play in gravity? We do not yet have complete answers to these questions. Nevertheless, physics has made tremendous progress in understanding the microscopic world and coming to terms with the variety of other ways that contradiction *seems* to appear.

13.1   Double Slit Experiment

The introduction to this chapter alluded to the duality paradox, or the dual wave-particle nature of matter. The most illuminating demonstration of this phenomenon is the double slit experiment, first performed by the British polymath Thomas Young in 1801, long before quantum mechanics was even a field. Young performed the experiment with light, and the same duality paradox has has since been demonstrated experimentally for electrons, atoms, and even highly complex molecules. The idea of the experiment is to shoot objects at a plate with two slits, beyond which lies a detector screen (figure 13.1).

Consider firing objects of two types at the screen: macroscopic and wavelike objects. If you fire a big, macroscopic object like a paint ball at the

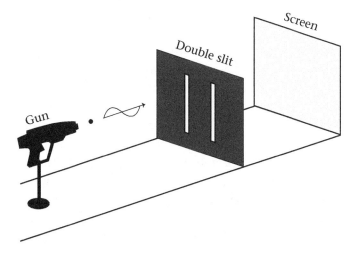

Figure 13.1
Setup of the double-slit experiment.

screen, then you can follow its trajectory with your eye, or maybe a camera if it moves faster than the eye can follow. It will travel through one of the slits and hit the detector screen. (We won't count the times when the paint ball hits the plate with the two slits on it and never makes it to the screen.) On the other hand, a wave-like object, like a water wave, has parts that will go through both slits and that will combine to produce a distinct pattern of crests and troughs on the detector screen.

The key point is that it seems like we have a device that can clearly discriminate between things that behave like particles (the paint balls) and things that behave like waves. So, what happens when you fire something *microscopic* at the screen, like an electron?

Claim 1

The screen distinguishes particles and waves, so it forces even microscopic objects to have a definite particle-like or wave-like character. Ergo, a quantum object can behave like either a particle or a wave, but not like both at the same time.

Claim 2

Quantum objects have inherently particle-like and wave-like properties. There must be some way that the double slit experiment demonstrates both of these sets of properties at the same time.

Discussion and Resolution

Claim 2 is correct. Here, the setup and results of the experiment are described for macroscopic objects (paint balls), and then for waves (water), and finally for microscopic objects (electrons).

We begin with the macroscopic case and use paint balls as an example. Suppose we fire paint balls through the pair of slits. On the other side of the slits is a detector screen, where the paint balls leave marks. The intensity of the color on the screen maps out the distribution of the paint ball arrivals. The balls are shot erratically, passing through the slits at various angles.

If we leave one slit closed for the moment, then we expect that there will be a single strip of paint marks on the detector screen, with dimensions proportional to the width of the open slit. The paint marks will be distributed probabilistically according to a curve $P_1(x)$, which gives the probability of a ball landing at a point x.

Similarly, if we close the first slit and shoot paint balls through the second one, this will generate a distribution, $P_2(x)$. Finally, in the case where both slits are opened, sometimes the balls would pass through the first slit and sometimes they would pass through the second one. If the paint balls are shot through the slits randomly, then we expect that the resulting distribution on the screen would be a sum of the previous two distributions:

$$P_{12}(x) = \frac{1}{2}\,(P_1(x) + P_2(x)) \tag{13.1}$$

$P_1(x)$, P_2, and $P_{12}(x)$ are probability distributions, which can be estimated from the experiment as follows. If you break the wall into small slices of size δx, you can approximate $P(x)$ by measuring the number of balls landing in the range $(x, x + \delta x)$, divided by the total number of balls. This distribution measures the fraction of balls that have landed in the range $(x, x + \delta x)$ for every slice of the wall. This counting experiment produces a histogram and will approximate the true probability distribution more closely as the number of balls fired increases. Figure 13.2 shows how a pattern might emerge over time as the balls are fired at two open slits.

We now conduct the same experiment using water waves instead of paint balls. Imagine half-submerging our experimental setup in water and replacing the paint ball gun with a source of turbulence that causes the water to ripple. The screen is equipped to measure the intensity of the waves striking it; the greater the intensity, the more the screen lights up.

If we leave only the first slit open, the distribution $P_1(x)$ will be similar to that of the paint ball case. We will see what is more or less a single strip of light on the screen, with intensity peaking at the center of the strip and tapering off to the sides. The result is the same if we close the first slit and open the second one to generate the distribution $P_2(x)$. But when we keep both slits open, we do not see two strips on the screen. The water waves passing through the two slits will constructively and destructively interfere, creating an *interference pattern* (figure 13.3). If you ever played with water in the bathtub as a child, then you will know that two wave crests form an even bigger crest when they meet, and that if a crest meets a trough, they cancel each other out to form a flat surface. This is what we mean by *constructive* and *destructive* interference, respectively.

Due to interference, instead of $P_{12}(x) = \frac{1}{2}\,(P_1(x) + P_2(x))$, there is another term that depends on the distance between the two slits and the angle of the

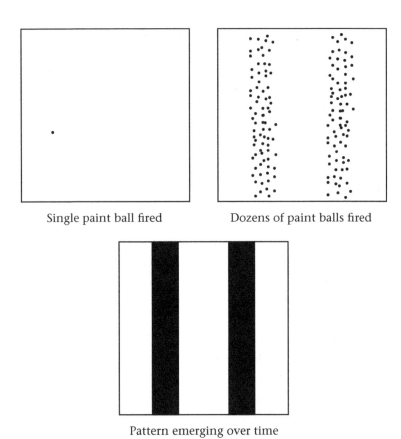

Single paint ball fired   Dozens of paint balls fired

Pattern emerging over time

Figure 13.2
Pattern on the screen as more and more paint balls are fired.

waves with respect to the straight-through path. The interference pattern creates what are known as *fringes*, which are the maxima in the distribution of intensity appearing on the screen. These fringes appear as a pattern of bright and dark bands on the detector screen.

In the third variation of this experiment, we fire electrons. Things are about to get strange. The experiment is run by shooting electrons through the slits. The electrons register on the screen as specks of light. The gun in figure 13.1 is now an *electron* gun, firing an electron beam that is so weak as to shoot only one electron at a time. If the electrons fire one at a time, just like the paint balls, then we should expect to see two strips emerge over time, just like the two strips of paint markings in figure 13.2, right?

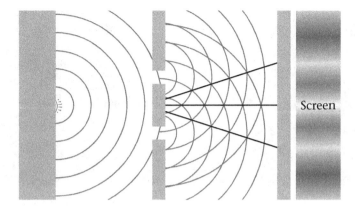

Figure 13.3
Interference pattern created by water waves passing through slits.

No, in fact that is not what happens. If only one slit is open, the specks of light on the screen form a single strip, consistent with the distribution created by the paint balls fired through one slit. But if both slits are open, the specks of light on the screen eventually form an interference pattern over time, as in figure 13.4, consistent with the distribution created by the water waves rippling through the two slits!

While we continue to shoot single electrons at a time through two slits, the distribution on the screen is not $P_{12}(x) = \frac{1}{2}[P_1(x) + P_2(x)]$. Rather, as we shoot more and more electrons, the light distribution on the screen shows an interference pattern, as seen when we used water waves. These *single electrons* are behaving like waves interfering with themselves.

If electrons behaved purely like particles, any fired electron would go through one slit or the other, like a paint ball. There would be no interference, and the pattern emerging on the screen over time would look like figure 13.2. But the double slit experiment shows this understanding to be incorrect. Each electron behaves as if it went through both slits simultaneously on its way to the screen, like a water wave. In other words, the electrons can exhibit particle-like properties (they appear as individual specks of light on the screen) and wave-like properties (they interfere with themselves as they propagate to the screen) at the same time.

Something interesting happens if you try to check which slit the electron goes through, however. Consider, for example, the Feynman light microscope experiment. In this experiment, light is shined on both of the slits,

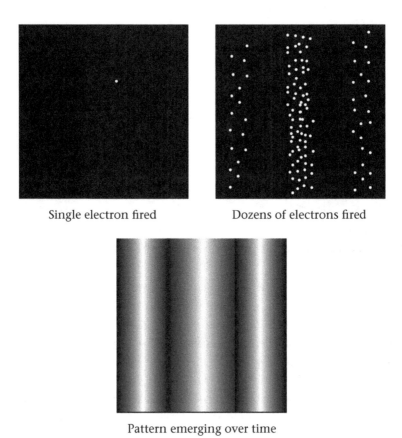

Single electron fired   Dozens of electrons fired

Pattern emerging over time

Figure 13.4
Pattern on the screen as more and more electrons are fired.

and a photodectector (a fancy word for what is essentially a camera) is placed by the slits. The photodetector can detect light that bounces off an electron as it travels through one of the slits, so in this way, one can deduce which slit each individual electron goes through. Curiously, this observation causes the electrons to stop behaving like waves, in that now the two-strip (paint ball) pattern is what appears on the screen, as shown in figure 13.2.

One way to understand the physics at play is to appeal to the wave function that we discussed in the introduction of this chapter. In the double slit experiment, when an electron "behaves like a wave," its wave function is spread out, and it probes the possible paths through the slits much as a water wave does. Shining light on the slits causes the electron to interact with this light, which in turn influences its wave function so it's no longer spread out in the same way. To use the correct terminology, we say that the electron is no longer *coherent* after interacting with the light; in other words, it can no longer interfere with itself, so the interference pattern goes away.

13.2 Schrödinger's Cat

At the start of this chapter, we introduced the concept of a wave function—the mathematical object that describes quantum systems. Born's rule says that the wave function calculates the probabilities for possible outcomes to be observed when a quantum system is measured. For example, the position wave function of an electron encodes the probabilities that it will be found at various locations when measured. How this mathematical object called the wave function corresponds to reality, as well as whether reality is actually probabilistic, are the central questions addressed by various interpretations of quantum mechanics.

Currently, the most popular of those interpretations is the Copenhagen interpretation, which was developed in the mid-1920s by Niels Bohr and Werner Heisenberg while they were working together in Copenhagen (although the two never entirely agreed). In this view, the properties of objects, including location, are not definite until they are measured. If objects aren't real without definite properties, then observation itself is what makes objects real. In the Copenhagen view, questions about the state of an object—like "Where is it?"—are meaningless before measurement. Measurement causes a phenomenon known as *wave function collapse*, whereby the object is forced into one of the states allowed by its wave function. Before wave function collapse, the object exists in a *coherent superposition* of all the quantum states allowed by the wave function.

In 1935, the Austrian physicist and Nobel laureate Erwin Schrödinger proposed a thought experiment as a critical commentary on the Copenhagen interpretation of quantum mechanics. Schrödinger's cat has become the most famous paradox of quantum mechanics in popular culture and provides a useful context for understanding the different interpretations of quantum mechanics.

The setup is shown in figure 13.5. A cat is placed in a box. This box has a radioactive atom that, being a quantum mechanical object, has a 50 percent probability of undergoing decay by a certain time. This means that it has two possible quantum states after the prescribed amount of time, decayed and nondecayed, and the atom's wave function predicts equal probabilities of

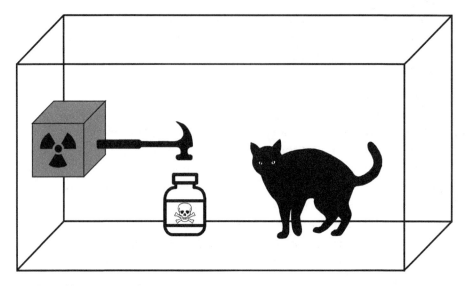

Figure 13.5
Schrödinger's cat.

observing each. A Geiger counter, which detects radioactivity, is connected to a hammer poised over a box of cyanide. If and when the atom decays, it triggers the detector, causing the hammer to break the box of cyanide, killing the cat.

Under the Copenhagen interpretation, after the prescribed amount of time, the radioactive atom exists in a superposition of decayed and non-decayed states. By extension, the cat similarly exists in a superposition of death and life states—simultaneously half dead *and* half alive.

Claim 1
The cat is in a superposition of states, both dead and alive.

Claim 2
The cat is either alive or dead, but not both.

Discussion and Resolution
Claim 2 is correct. This paradox explores how and whether a particle in superposition, through a series of Rube Goldberg–like dependencies, can cause a macroscopic system to exist in superposition. We've learned that

in the Copenhagen interpretation, measurement forces an object in super-position into a single state allowed by its wave function. But what does *measurement* actually mean here? This question is complicated, but generally, any interaction a quantum system has with its surrounding environment, with or without deliberate cause and conscious knowledge, constitutes a measurement.

Cats interact with their environments in many ways. For example, they emit thermal radiation, which interacts with the surrounding air molecules. This interaction occurs regardless of the presence of a conscious observer, and it is but one of the large number of interactions that cause the state of the radioactive atom to be measured. In addition, the Geiger counter interacts with the atom in such a way as to cause measurement.

To use the correct technical language, we say that the radioactive atom is in a *coherent* superposition before measurement. If it were possible to perfectly isolate the atom from everything else, then it would truly exist as a superposition of being decayed and nondecayed. We could leverage this coherence to perform further experiments that were really quantum; for example, these two states could be made to interfere with each other, much as the electron interferes with itself in the double slit experiment. In Schrödinger's thought experiment, however, the radioactive atom is any-thing but isolated from the environment. Environmental interactions de-grade its coherence. We say that the radioactive atom *decoheres*.

A resolution consistent with the Copenhagen view is that the atom never has a chance to exist in superposition. As an element of the macroscopic system that includes the cat and Geiger counter, it is measured, or equiva-lently, it is not at all robust with respect to decoherence. That measurement causes the wave function to collapse into a state of decay or nondecay. The cat is either dead or alive before anyone opens the box to determine which. A human opening the box will find the cat dead with 50 percent probability, and alive with 50 percent probability, but the cat is never both alive and dead. This uncertainty is purely classical, in the same way that you have a 1 in 52 chance of drawing any particular card from a fair deck of playing cards whose sorting order is unknown to you. It's an *incoherent superposition*.

Another popular interpretation of quantum mechanics is the Many Worlds view, formally originated by the American physicist Hugh Everett in 1957, though Schrödinger had made prior reference to the basic idea.

This view holds that measurement phenomena cause the universe to split into many noncommunicating branches, each with a history reflecting one of the possible outcomes—that an astonishingly large and ever-expanding collection of parallel universes constitutes a greater multiverse. This interpretation rejects the idea of wave function collapse, asserting instead that a wave function decoheres into sets of branches that continue to exist in their own right. According to the Many Worlds interpretation, opening the box and finding the cat dead versus finding the cat alive correspond with different branches of the same wave function.

The strange result of the double slit experiment also can be explained by Many Worlds. The electron, as it travels toward the detector screen, first exists as a superposition of having traveled through the first slit and having traveled through the second slit. When the electron hits the screen, however, the total wave function of the screen and electron branches according to the possible locations where the electron could hit the screen, and we, in the laboratory, see one such branch. In this way, you can think of the different states of motion that make up the electron's wave function before it is measured by the screen as *prebranches*. David Deutsch, a strong advocate of the Many Worlds theory, explains that the interference pattern created by firing single electrons through the double slit is created by what will become the various versions of that electron, in different universes (i.e., branches), interfering with each other. Deutsch writes in his book *The Fabric of Reality*:

> The quantum theory of parallel universes is not the problem, it is the solution. It is not some troublesome, optional interpretation emerging from arcane theoretical considerations. It is the explanation—the only one that is tenable—of a remarkable and counterintuitive reality.

The Many Worlds view indeed has some fascinating implications. It implies there are countless versions of *you* throughout the multiplicity of universes—perhaps even versions in which you're a billionaire or an Olympian. It implies, to a high degree of probability, universes with histories in which your parents never met. It implies the existence of universes with exceedingly bizarre histories.

In some sense, the Many Worlds view even supports the idea of an immortal version of *you*. Consider a quantum gun that fires with a 50 percent probability, as governed by the measurement of some particle in a quantum state. (Alternatively, it's as if you're the cat in the box.) According to the Many Worlds view, when you fire the gun at yourself, the universe splits in two: one in which you die, and one in which you survive. The surviving version of you could fire the gun again, creating a new dead you and a new surviving you. Every version of you that survives, by firing once more, will create a new version of you that dies and one that survives. There will *always* be a version of you that escapes suicide by sheer luck.

There are other interpretations of quantum mechanics, too. A large class of interpretations, for example, can be grouped together under the banner of *hidden variables*. Unlike the Copenhagen interpretation and Many Worlds, these interpretations maintain that reality is not actually probabilistic. Instead, quantum mechanics only appears probabilistic because it is an incomplete description of reality. In relation to quantum mechanics, therefore, there should be a set of underlying degrees of freedom, the hidden variables, which deterministically predict the outcomes of experiments that are probabilistic according to minimal quantum theory. Different interpretations offer different explanations for what these hidden variables might be and how they might work; however, we will see in section 13.4 on the Einstein-Podolsky-Rosen Paradox that some aspects of hidden variable interpretations can actually be constrained experimentally. While the paradox of Schrödinger's cat may be resolved within the various interpretations, debate continues as to which interpretation describes reality.

On the practical side of things, coherent quantum states are difficult to create. Small quantum systems are inherently susceptible to interactions with their environments that lead to decoherence, hence the difficulty in creating technologies like quantum computers. At best, advanced labs have succeeded in placing a few well-isolated atoms into a coherent quantum state. Quantum computers require being able to put a large number of such small systems into a coherent quantum state, and furthermore being able to reliably manipulate that state.

The rewards of building a quantum computer will be well worth the trouble, however. Because the logic components of quantum computers are made up of quantum systems, algorithms that run on a quantum computer can exploit quantum resources, like coherence and interference. Another important resource for quantum computation is entanglement, which was mentioned briefly in the introduction and will appear in more detail in the last paradox in this chapter. These resources are unavailable to classical computers, and they can be leveraged to solve problems that otherwise would be inefficient to solve classically.

Quantum computers aren't miracle machines, however. The problems that quantum computers can solve efficiently are limited in nature, typically to the study of relationships between solutions to problems rather than the values of solutions. Nonetheless, scientists hope that by learning how to overcome quantum decoherence in controlled settings, they will pave the way for technology to harness life-altering (or at the very least, science-altering) processing power.

13.3 Turing Paradox (Quantum Zeno Effect)

Until now, we have given little consideration, other than implicitly, to the fact that quantum systems can change in time just as classical systems do. The study of the change of quantum systems is called *quantum dynamics*.

When a quantum system is measured, it is "forced" into one of the states allowed by its wave function. In this discussion, we have learned there's a variety of interpretations of this event—the collapse of the wave function, the branching of the wave function, and other explanations—but what remains consistent is that a definite outcome is measured from a set of possibilities governed by the wave function.

The wave function and the probabilities that it governs evolve in time. To dig a bit deeper, let's consider a simple model. Many particles and atoms have an intrinsic *magnetic moment*; that is, they generate a small magnetic field and behave like little magnets when probed. It turns out that the orientation of an atom's magnetic moment is quantized, and for many atoms, like certain isotopes of silver, the magnetic moment can point either up or down along a given axis. The magnetic moment of such an atom therefore has two quantum states: up and down.

Suppose that we have such an atom and that we measure the state of its magnetic moment along a given axis, which we label the *x*-axis—and find that it is in the up state. Due to our measurement, the atom is forced into this state. Next, suppose that we turn on an electromagnet that produces a magnetic field in a perpendicular direction, say along the *z*-axis. It turns out that this causes the atom's state to start oscillating between pointing up and down, and at any point in between, the magnetic moment is in a superposition of pointing up and down. Neat.

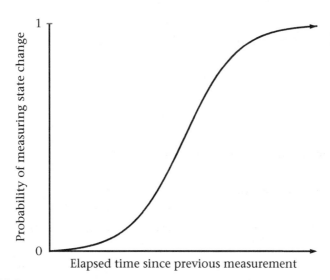

Figure 13.6
Probability of measuring a state change as time passes since the previous measurement.

Now suppose that we measure the state again after some amount of time. Figure 13.6 approximates the relationship between elapsed time between measurements and the probability of measuring that the atom's state flipped to the down state over half of a full oscillation. This plot indicates that, at least for the range of times shown, we are more likely to measure down next time the longer we wait.

But wait a minute. Notice that the probability of measuring a state change approaches zero as we reduce the time between measurements. If we make another measurement quickly after the first, the atom doesn't have much time to superpose. It's likely to be measured in the same state as last time. The more frequently it is observed within a given time frame, the less likely it is to change states in that time frame.

This is the *quantum Zeno effect*, proposed by Alan Turing and named for its similarity to a paradox described in chapter 2: Zeno's Arrow Paradox, which purports to show the impossibility of motion by dissecting time into a set of instants.

The saying goes that "a watched pot never boils." Of course, this isn't really true; it's a statement about the psychology of impatience. But is there some sense in which the proverb is actually a fundamental fact of the universe? Does a watched atom never change states? Does watching a quantum system prevent it from evolving in time?

Claim 1
Watching an object cannot affect its natural evolution in time.

Claim 2
Watching an object can affect its evolution in time.

Discussion and Resolution
Claim 2 is correct. It is important to realize that the quantum Zeno effect doesn't stop time. It merely affects the evolution of a quantum state as time passes.

Notice in figure 13.6 that the probability of measuring a state change approaches 1 if we wait a certain amount of time between measurements in our magnetic moment example. This implies that by waiting for an appropriate amount of time, we can increase the probability of forcing the atom to change states. This is called the *quantum anti-Zeno effect*. By a process of timing measurements, objects can be manipulated to evolve in a particular way in time.

Both the quantum Zeno effect and the quantum anti-Zeno effect have been observed experimentally numerous times. To demonstrate the former, in 2015, a group of scientists at Cornell University laser-cooled a collection of rubidium atoms to just above absolute zero in a vacuum chamber. Ordinarily, the rubidium atoms would form a crystalline lattice and tunnel around the lattice. Experimenters shined an imaging laser on the lattice to measure the degree of tunneling. They found that the brighter the imaging light—that is, the more the lattice was "watched"—the less tunneling took place. What's remarkable about the experiment is that the quantum property under observation was less esoteric than the orbital radius or magnetic moment orientation; it was location in space. Watching the rubidium atoms kept them in place.

Fans of the television series *Doctor Who* might be reminded of the Weeping Angels, a villainous race that can move quickly and stealthily, so long as no one is watching. The moment someone looks, they freeze and turn to stone, "quantum-locked," until no longer under observation. At the Cornell lab, the rubidium atoms behaved like Weeping Angels.

The quantum Zeno effect may have important implications in biology, particularly for certain migratory birds. Scientists have not yet fully understood how birds use the Earth's magnetic field to navigate, but there are some well-developed theories. One suggests that the magnetic field may influence chemical reactions taking place in bird retinas. These chemical reactions would ordinarily happen too fast for the magnetic field to matter. But the reactions may depend on the state of electrons. It has been hypothesized that these electrons maintain their state long enough to enable magnetic field influence, as a result of the quantum Zeno effect within bird eyes.

In principle, with an established progression of quantum states with known time scales, it would be possible for a quantum system to be walked through a sequence of states if interactions with the system were timed to the right cadence. The quantum Zeno effect is an important tool in the field of quantum computing, and it will be a candidate for storing quantum states and preventing the decoherence of systems.

13.4 Einstein-Podolsky-Rosen Paradox

The last big quantum idea that we will discuss in this section is entanglement. When you have a quantum system made of several components, like a collection of particles, the individual components can develop correlations among themselves that are so strong that measuring the state of one component can influence the state of the others. This is a characteristic feature of entangled degrees of freedom.

Einstein in particular did not like entanglement. In his view, it flew in the face of a principle he called *physical reality*. Together with two of his colleagues, Boris Podolsky and Nathan Rosen, he articulated his objection through the Einstein-Podolsky-Rosen Paradox.

To begin, we should talk a bit more about entanglement, and to that end, an example is useful. Recall from section 13.3 on the Turing Paradox that some particles have an intrinsic magnetic moment. Just as a particle can have a certain electric charge or a certain mass, some particles have a tiny amount of intrinsic magnetism, as if they were tiny bar magnets. Furthermore, recall that the direction in which the magnetic moment can point along a given axis is quantized. For example, both an electron and its antimatter counterpart (called a *positron*) have a magnetic moment that can point either up or down along a given axis. If we pick an axis and call it the *x*-axis, then we say that an electron or a positron has two states: up along the *x*-axis, and down along the *x*-axis.

You may have heard of a type of particle called the *neutral pion*. This is a composite particle that is made up of the same stuff (quarks) as protons and neutrons. In particular, the neutral pion is unstable, and it can decay into an electron and a positron. Denoting the electron by e^-, the positron by e^+, and the neutral pion by π^0, we can write out this process as

$$\pi^0 \longrightarrow e^- + e^+ \tag{13.2}$$

A neutral pion has no intrinsic magnetic moment. Due to the way that intrinsic magnetic moment arises, a consequence is that the electron and positron have to be produced with their magnetic moments antialigned; they have to be in opposite up/down states. (Heuristically, the electron's and positron's magnetic moments have to cancel out to zero because the pion had no magnetic moment in the first place.) When the pion decays, however, whether the electron is in the up state and the positron is in the down state, or vice versa, is not determined. One could schematically write out this superposition as

$(e^- \downarrow$ along x and $e^+ \uparrow$ along $x) + (e^- \uparrow$ along x and $e^+ \downarrow$ along $x)$

Notice that if you measure one particle (say the electron), then you immediately know what the state of the positron must be; it's the opposite. You never actually need to do anything to the positron to reach this conclusion. This is an example of entanglement. Here, the electron and the positron are entangled with each other.

Trouble is already brewing, especially if we think relativistically. Our conclusion here may seem mild if the electron and positron are in the same laboratory. But what if they are at opposite sides of the building, or opposite sides of the planet, or even opposite sides of the visible universe? It seems troublesome that by measuring one particle, we immediately influence the state of the other. It would seem that we can influence a faraway system faster than the speed of light, or in other words, *nonlocally*.

Einstein, Podolsky, and Rosen sharpened the issue even further. Notice that our choice to measure the direction of the electron's and positron's magnetic moments along the *x*-axis was arbitrary. We could just as well have chosen to measure along an orthogonal axis, like the *z*-axis. In this case, we would have arrived at the same conclusion regarding the joint state of the electron and positron:

$$(e^- \downarrow \text{ along } z \text{ and } e^+ \uparrow \text{ along } z) + (e^- \uparrow \text{ along } z \text{ and } e^+ \downarrow \text{ along } z)$$

Crucially, in the same way that position and momentum are governed by an uncertainty principle, wherein both cannot be known to arbitrarily fine accuracy simultaneously, so too are the orientations of magnetic moments along the x-axis and the z-axis. Recall from the Turing Paradox, for instance, that the state "up along x" is an equal superposition of "up along z" and "down along z." If you measure an electron and find that its magnetic moment points up along the x-axis, then a measurement along the z-axis (as up or down) will have a totally random outcome.

To Einstein, the *physical reality* of a quantity meant that you must be able to predict its value without disturbing it. In that sense, the orientations of an electron's magnetic moment along the x-axis and along the z-axis individually have physical reality, but they do not have *simultaneous* physical reality. The problem with the entangled electron and positron is that the mere possibility of choosing to measure the electron—and only the electron—along the x-axis or the z-axis causes the magnetic moment of the positron to have a definite value, but along mutually incompatible directions. Because both definite outcomes for the positron are possible and predictable, depending only on an arbitrary choice of measurement by the experimenter, even when the positron is totally isolated and causally disconnected from the experimenter, Einstein concluded that the orientation of the positron's magnetic moment along *both* the x- and z-axes must have simultaneous physical reality. This is the only possibility if everything, including entanglement, is local in the relativistic sense. There is the paradox: This simultaneous physical reality is not allowed by the rules of quantum mechanics that we have seen so far.

Claim 1

Quantum mechanics is an incomplete theory. It must be that there is some set of underlying degrees of freedom, which are not part of quantum mechanics, that determines the probabilistic outcomes of mutually uncertain measurements. Then, mutually uncertain quantities, like position and momentum or the orientation of magnetic moment along the x- and z-axes, have simultaneous physical reality. This is the claim that was made by Einstein, Podolsky, and Rosen.

Claim 2

Quantum mechanics has nonlocal correlations that are so strong that measurements made on one part of a system can influence the other parts.

Discussion and Resolution

As far as is known, Claim 2 is the correct resolution, at least if we further qualify Claim 1. Einstein, Podolsky, and Rosen hoped that there was some set of underlying degrees of freedom called *hidden variables* that would resolve the probabilistic character of quantum mechanics and that, crucially, were local. Then, it was the experiment of the physicist John Stewart Bell that falsified the claim advocating for local hidden variables.

Bell's original experiment is actually quite complicated, so here we will describe a slightly simpler variation that was proposed by John Clauser, Michael Horne, Abner Shimony, and Richard Holt, abbreviated by CHSH. Suppose that Alice and Bob are in a lab and that they possess an apparatus that generates pairs of electrons and positrons for them. Alice and Bob each have a device that they can individually configure to measure the magnetic moment of the particle that they receive along two possible axes. Each time they receive a particle, Alice can select to measure along axis a or axis a', and Bob can select to measure along axis b or axis b', and they can make these choices independently.

First, Alice and Bob set their detectors to the a and b settings. Each time they receive a particle and perform a measurement, Alice's detector displays $A = 1$ if the particle's magnetic moment was measured to point up along the a-axis, and $A = -1$ if it pointed down along the a-axis. Similarly, Bob's detector displays either $B = 1$ or $B = -1$. Every time they perform a measurement, they multiply their detector readings together to compute the value of AB. After performing many measurements, they can then get a good estimate for the average value of AB, denoted by $\langle AB \rangle$, which they then record in their lab notebook.

Next, they repeat the protocol, but this time with their detectors configured to measure along the a'- and b-axes. In this way, they obtain an estimate for the average $\langle A'B \rangle$. They then repeat the protocol two more times, with their detectors set to a and b', then a' and b', to estimate the averages $\langle AB' \rangle$ and $\langle A'B' \rangle$.

Finally, Alice and Bob consult their notebook and compute the value of

$$C = \langle AB \rangle + \langle A'B \rangle + \langle AB' \rangle - \langle A'B' \rangle . \qquad (13.3)$$

Crucially, CHSH (and also Bell, in his version of the experiment) demonstrated that in *any* theory where quantum mechanics is explained by local hidden variables, the value of C is always between -2 and 2, regardless of the state in which the electron and positron are prepared and what detector settings Alice and Bob choose. However, when this experiment is performed using entangled electrons and positrons, then by carefully choosing the detector settings, one finds that C can be as large as $2\sqrt{2}$ or as small as $-2\sqrt{2}$, which is well in violation of CHSH's and Bell's bound.

The conclusion is that there are no local hidden variables that can explain the probabilistic nature of quantum mechanics. It seems that we just have to cope with probabilistic outcomes and nonlocal correlations.

Speaking of this, how do we reconcile the nonlocality of entanglement with relativity? While entanglement means nonlocal correlations, it turns out that you cannot exploit these correlations to violate causality. Entanglement alone does not let you communicate faster than the speed of light. For example, suppose that you hold half of an entangled pair of particles whose other half is at the other side of the visible universe. Then, there's a precise sense in which you can't do anything useful with this half of the entangled pair. After all, you have no way of knowing that any interesting manipulations were performed on the other half unless someone tells you about them. In order to tell you, the other party has to communicate with you classically, which can be done only using signals that travel at or below the speed of light. Therefore, while entanglement is nonlocal, it's not strong enough to break causality.

In closing, Bell's and CHSH's experiments are by no means the end of the story. Hidden variables may still be a viable means of explaining the indeterminacy of quantum mechanics—they just cannot be local hidden variables of the type originally envisioned by proponents of the theory. Theoretical physics abounds with more exotic proposals, too. For instance, one suggestion is that there could be a sense in which any pair of entangled particles are connected by a *quantum wormhole*. Then, entangled particles would be right next to each other in space and time, which would explain their strong correlations that seem nonlocal. Either way, entanglement is an important feature of quantum mechanics that is central to the developments of the theory, both past and future.

14 Invented or Discovered?

Pure mathematics is, in its way, the poetry of logical ideas.
—Albert Einstein

Some paradoxes in this book have pointed to restrictions imposed by mathematics, or the limitations of mathematics itself. Arrow's Impossibility Theorem proved that no voting system can meet all reasonable criteria of fairness. Cantor's Paradox showed us that some things, like infinities, can't all be gathered into a set. A stunning blow was delivered by Gödel's Incompleteness Theorems, which established that every axiomatic system containing basic arithmetic is either incomplete or inconsistent.

Given these and other limitations, how well can we trust mathematics? Central to that question is the origin of mathematics itself. Is mathematics a collection of discoveries of absolute truth, or an intricate invention by man? Simply put, is mathematics invented or discovered? This question has fascinated a great many thinkers.

An *invention* is anything created by an intelligence that is not an inherent feature of nature. A *discovery* is an inherent feature of nature that an intelligence integrates into its body of knowledge. Which is mathematics? On the one hand, mathematics is a tool with practical applications. Mathematics provides a foundation for engineering, building, and technology design, all of which help us adapt to our environment and flourish. It has its own language, a set of symbols and grammar, created by humans. Viewed through this lens, it would seem proper to call mathematics an invention, like any other that has benefited human life. On the other hand, how could mathematics help mankind cope with nature if it didn't have a basis in reality? Mathematics is not a set of artificial rules in an artificial context, like the game of Monopoly. Mathematics has a correspondence to objective truth, and viewed through this lens, it seems better described as discovery.

So, which is it? Invention or discovery? This classification problem is a paradox all its own. To answer the question, let's create a framework for understanding how concepts correspond to reality. We'll need to distinguish between four things: real phenomena, concepts, words, and real referents.

- A *real phenomenon* is a natural feature of reality, of which there can be many instances. For example, the existence of giant burning balls of gas in space is a real phenomenon. There are many instances of this phenomenon around us: individual burning balls of gas throughout the universe.
- A *concept*, as defined by Ayn Rand, is a mental integration of two or more instances that are united by a specific definition. In the example here, the definition is the idea expressed by the words *giant burning ball of gas in space*.
- The *word* is the linguistic tag that we give a concept, which helps us communicate and mentally organize the concept. In this example, the word is *star*.
- The *real referent* of a concept is defined here as the real phenomenon to which the concept corresponds. The real referent of the concept of *star* is the factual existence of burning balls of gas in space.

This framework applies to intangible real phenomena as well, like properties of objects. One example is motion. Objects change their positions over time. That is a real phenomenon. There are many instances of motion all around you. You can't touch any of those instances; you can touch a moving object, but you can't touch motion itself. Nonetheless, you can form a concept of motion and tag it with the word *motion*. The real referent of the concept of motion is the fact that things change their positions over time. That fact is independent of you or any conscious perceiver.

Properties of things are intangible, but they are real phenomena, so this framework still applies. Let's take a look at more subtle properties of things. Consider the types of chess pieces. Each side has pawns, knights, bishops, rooks, a queen, and a king. It's safe to say that together, the black pawns and the white pawns constitute all the pawns; every piece that is a black pawn, together with every piece that is a white pawn, constitute the totality of pawns. We can separate them into eight black pawns and eight white pawns on opposite sides of the board, or we can line them up next to each other and have sixteen pawns. When we line up all the pawns, we don't end up with any other pieces in the mix. A pawn is always a pawn.

The same series of statements can be said of any other type of piece, although the other pieces come in different numbers. In that sense, there is a real phenomenon going on here, intangible though it may be, and that phenomenon instantiates itself across all the different piece types. The instances of that phenomenon can be mentally integrated and united by a definition, forming a concept with a real referent.

That concept here is that of the *union*, which is described formally by the Axiom of Union in ZFC. There are nine axioms in ZFC, each of which has a real referent (though in some cases, it is harder to see how that's true). Together, these axioms establish the notion of a *set*. A set may be a mental abstraction, but it is conceived in direct correspondence to real phenomena.

Not every mathematical concept has an immediate real referent. In fact, most do not. This does not make mathematics any less grounded in reality; a wall can be considered solidly grounded even if most of its bricks have no direct connection with the soil. Most concepts are built from conceptual precursors. For example, one must first have a notion of a set before having the notions of *bijection* and *positional ordering*; and these are respective conceptual precursors to *cardinality* and *ordinality*, which are conceptual precursors to two kinds of *number*. But every mathematical concept, however complex, exists on a ladder, the lower rungs of which have real referents. Even the imaginary number *i* can be found on that ladder, divorced from reality as it may seem.

The symbols and grammar used in the language of mathematics are certainly invented. But the concepts expressed by this language all trace back to real referents, either directly or through a ladder. A system of symbols, grammar, axioms, and rules of inference together comprise a formal system of mathematics, as discussed in chapter 7. Some aspects of a formal system are invented, while others take root in discovery. Theorems emerge from formal systems. Theorems are mathematical statements that can be proved within a formal system from the axioms using the rules of inference. When a formal system is established, the provability of certain theorems within that system becomes a feature of nature. Theorems, therefore, are best described as discoveries.

In summary:

- The most fundamental concepts, or *axioms*, of a formal system are grounded in real phenomena that intelligence discovers. In a sense, concepts are cognitive inventions grounded in discovery.
- Other more advanced concepts can be built from these fundamental concepts, but ultimately every mathematical concept is grounded in real phenomena.
- The symbols and grammar of any formal system are invented to suit the concepts. Once a formal system is established, provable theorems emerge, available to discover.

Another Approach

Among my friends is a talented educator in the field of mathematics: Grant Sanderson, creator of the popular 3blue1brown YouTube channel on mathematics. I invited Grant to share his approach to the "invented or discovered" question and am grateful for his contribution of the following original material.

14.1 Essay and Poem by Grant Sanderson

Many words in other languages have no direct English translation. The Finnish word *poronkusema* means "the distance a reindeer can comfortably travel before taking a break." The Inuit word *iktsuarpok* means "the feeling of anticipation when you're expecting someone that leads you to constantly check to see if he or she is coming." The German word *Kummerspeck* translates to "grief bacon" and refers to the the weight one puts on after overeating during an emotional time.

Would you say these words are invented or discovered?

On the one hand, they seem clearly invented. Words like these are not unearthed on lost tablets or found by experiment. They exist in language because humans decided they should, typically as a result of many small choices through the evolution of a culture. It would seem, then, that the presence of these words would depend entirely on the peculiarities of people and their whims.

On the other hand, these words represent observations of the human condition. In a society without emotional overeating, the word *Kummerspeck* would have no reason to come about. If an alien visiting Earth were to learn such a word, the alien would come away with some information about our world, our culture, and our people. Even if the symbols, sounds, and definitions characterizing these words are made up, their existence is somehow related to discovery.

If you're at all on the fence as to whether the ideas behind words like *Kummerspeck* are invented or discovered, you can understand why thinkers have long pondered the same question as it pertains to mathematics. There are two ways the former question parallels the latter.

First, for each equation, theorem, or construct you encounter in math, some individual or group made the decision to articulate it and put it into common usage. In this sense, as with words, they feel invented. At the same time, many of these facts have a dual feeling of being discovered, resting on some truth of the world that would exist independent of observation. Emotional overeating happens whether or not there is a word for it, and the ratio of a circle's circumference to its diameter is always $\pi = 3.14159\ldots$ even if no one computes it.

The second, more substantive side of the parallel has to do with the context in which each feels invented versus discovered. Loosely, both for words and mathematical facts, the more you ask of something, "What is it, really?" the more it feels invented; and the more you ask, "How does this apply to the world?" the more it feels discovered.

Let's spell this out a bit more in the case of math, with the Pythagorean theorem as an example. In a right triangle (a triangle containing a 90° angle), the sum of the squares of the lengths of the legs (the two smaller sides) equals the square of the hypotenuse (as shown in figure 14.1).

This certainly feels discovered. Upon learning this, you might say to yourself: "Wait, really?" You can draw right triangles and verify the theorem with a ruler. You could imagine an intelligent civilization in another galaxy stumbling upon the same fact.

Something peculiar happens as you unpack the mathematics behind this fact. You must ask: What exactly is a *line*? What is meant by *length*? What does it mean for an angle to have 90 *degrees*? Sure, you can show what a line looks like by drawing it, and you can measure lengths or angles with a ruler or protractor. But mathematicians strive to rely as little as possible on sensory perception. Mathematicians begin with a small number of primitive objects and assumptions about their properties. From these, they make logical deductions and further constructions.

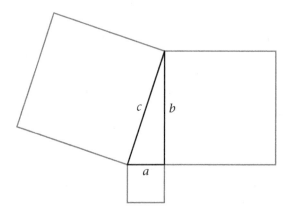

Figure 14.1
The Pythagorean theorem states $a^2 + b^2 = c^2$.

Sets are a common building block. They aren't the only building block; modern mathematics includes alternative frameworks with other primitive objects, but set theory is the most mainstream. It's possible to fully construct the concepts of lines and two-dimensional space in terms of sets and nothing else. No drawing is necessary, nor is dependence on any conception of space or familiarity with spatial reasoning.

Here's just a little taste of how mathematical objects can be constructed from sets. Let's start with the empty set, {}. Right now, our mathematical universe includes nothing at all to work with, other than the notion of collections of elements. Since there's nothing in our universe to be contained in a set, all we have is {}. But with that, we can also have {{}}, the set containing the empty set. And this gives us a few more things, like {{}, {{}}}. It's confusing to read with all those brackets, so let's use names. Let 0 represent the empty set {}, and let 1 represent the set containing 0:

$$1 = \{0\} = \{\{\}\}$$

The set containing those two, $\{0, 1\}$, will be 2. Likewise, 3 is $\{0, 1, 2\}$, 4 is $\{0, 1, 2, 3\}$, and so on. In fact, let's say that this is how we define the natural numbers. This is an example of constructing something using only sets. It's a bit frightening to think of what a larger number like 100 represents when fully unpacked in bracket notation, but in principle, you can understand that it's turtles all the way down. There's no need for a construct more primitive than sets.

With some more work, you can define the ideas of functions and operations, like addition, subtraction, multiplication, and division. You can define fractions, and with some cleverness, you can define real numbers. With real numbers, you can get some notion of two-dimensional space as all pairs of real numbers. This is a notion that could be communicated to an intelligent being with no perception or conception of space, but only an ability to understand what you meant by *set* and *contain*.

Architects, engineers, and other craftsmen do well enough working with geometry without thinking of space as nested sets. Why do mathematicians bother with so much rigor when our shared intuition works well for practical purposes? It is because removing our reliance on intuition helps us discover truths contrary to that intuition. Physicists have historically benefited from this challenge. Counterintuitive discoveries made in the lab often have been describable with mathematics discovered decades or centuries ago. Rigorous mathematics also helps us understand how the basics of human life, like motion, are possible. The paradoxes that perplexed Zeno, for example, are explained entirely by mathematical rigors. The resolutions show that motion is in fact possible and that the real world is not static and timeless. Mathematical rigor provides a foundation for learning how the world really works.

Returning to our Pythagorean example: In any formalism of geometric concepts like two-dimensional space, lines and triangles, we have a choice as to how we define length. There are many internally consistent notions of length, called *norms*. One standard norm is the Euclidean norm, which essentially captures the Pythagorean theorem not by proof, but by definition.

Does that seem troubling? Before, the Pythagorean theorem felt like a discovered fact about the world. It could be proved logically, verified empirically, used practically. The Pythagorean theorem was "discovered" and put into practice long before humans had any notion of set theory or a formal underpinning for space as a composite of more primitive objects. If our Euclidean norm is Pythagorean in character simply by definition, have we lost something?

To appreciate the new role that the "discovery" of the Pythagorean theorem has in this setting, it's important to understand that alternative choices for how to define length also give rise to consistent mathematical theories, many of which can be useful in other contexts. There are some settings in which the Pythagorean theorem is not true, like on curved surfaces. Curved space is important to modern physics, so having a modified formalism is handy. Space can be envisioned in other ways that are internally consistent, but they do not align with the physical world.

So, if there are multiple internally consistent ways to define length, how do you know which one is right? Well, there is no right answer. The better framing is to ask which choice gives rise to a mathematical theory that is useful in a given setting. In the setting of flat space, where the Pythagorean theorem was "discovered," using that theorem to define length will give rise to a geometry that provides accurate predictions in that setting. Discoveries in math guide us in creating formalisms that will be useful.

The logical foundations of a mathematical theory are not the beginning of that theory, but an important middle step in its development. The theory begins with discoveries, potentially informal ones–observations about the world stumbled upon either in an empirical context or in a theoretical context of an adjacent theory. Those discoveries, like the Pythagorean theorem, inform the shaping of a more rigorous theory, which in turn creates new venues for discovery.

This cycle echoes through the history of math. Imaginary numbers arose as the residue of a method discovered by Gerolamo Cardano for solving cubic equations in the sixteenth century. Once the mathematical community accepted the square root of -1 as a legitimate concept, i opened the door for myriad discoveries in mathematics and other fields. These, in turn, led to the definition of new and more intricate constructs, like Riemann surfaces and analytic functions, which opened still more new doors.

The invention and discovery of mathematics have existed not in a philosopher's tug-of-war, but in a cycle of progress. It's not too different with words and language. Observations about the world, and particularly those significant to human life, give rise to the invention of words. Language forms around phenomena important enough to deserve conceptualization and designation by label. A language with a richer vocabulary can express subtler observations, which may merit the formation of yet-newer words. So the cycle continues.

I leave you with a poem, dear reader: an invention of mine, based on discovery.

Invented or Discovered?

A lurking question, old as Greece, does ask:
Is math invention or discovery?
The answer's both, but here the harder task
Is knowing how it switches. So you see,
Most truths are found with constructs, which in turn
Are built from truths like chickens and their eggs.
One needs to know, should this fact raise concern,
That constructs can be daft; just logic's dregs.
The truths discovered tell how to define
The structures that make math and world align.

Pythagoras's theorem about lengths,
Is one such truth that he (and others) found.
Observed, discovered, and through mental strength,
It was then proved with pictures that astound.
Then later mathematicians, minds alit,
Would formalize, as sets, both "length" and "space."
These constructs were defined so that by writ
Pythagoras's truth remains the case.
That is, this "theorem" now seems just defined.
As such, one might then lose his peace of mind.

So in this world where "space" is formalized,
Do all those pretty proofs now lose their charm?
Of course not! We all ought to realize
How "length" could change its meaning without harm
To mathematical consistency
In all the theory giving facts of space.
However, then these "facts" would hardly be
Related to the nature we embrace.
And hence, a truth has told how to define
A structure that makes math and world align.

Notation Guide

Relations

Symbol	Meaning
=	equals
≠	does not equal
>	strictly greater than
<	strictly less than
≥	greater than or equal to
≤	less than or equal to
≡	is defined as / is equivalent to
≻	is preferred to

Logic

Symbol	Meaning
∃	there exists
∄	there does not exist
∀	for all
¬	not / negation
∨	or / inclusive disjunction
∧	and / conjunction
→	implies / only if
←	is implied by / if
↔	is equivalent to / if and only if

Greek Letters

Not all Greek letters are used in this book, but all are listed here (in alphabetical order) for completeness.

Uppercase, lowercase	Name	Uppercase, lowercase	Name
A, α	alpha	N, ν	nu
B, β	beta	Ξ, ξ	xi
Γ, γ	gamma	O, o	omicron
Δ, δ	delta	Π, π	pi
E, ε	epsilon	P, ρ	rho
Z, ζ	zeta	Σ, σ	sigma
H, η	eta	T, τ	tau
Θ, θ	theta	Υ, υ	upsilon
I, ι	iota	Φ, ϕ	phi
K, κ	kappa	X, χ	chi
Λ, λ	lambda	Ψ, ψ	psi
M, μ	mu	Ω, ω	omega

Infinite Cardinals

The only Hebrew letter used in this book is *aleph*. Its symbol is $\aleph$. Aleph is used as a symbol for infinite cardinal numbers that can be well-ordered. The first infinite cardinal is $\aleph_0$, which reads as "aleph-naught," "aleph-null," or "aleph-zero" and represents the cardinality of the set of natural numbers. Aleph is not to be confused with the generic symbol for infinity, ∞.

The cardinality of the continuum can be expressed as the cardinality of the power set of the naturals, $2^{\aleph_0}$. It also has its own symbol, c.

Set Notation

A *set* is a collection of *elements*, which are members of the set. Sets are usually denoted by normal capital letters. Some special sets are associated with the following symbols:

Symbol	Meaning
$\mathbb{N}$	the natural numbers
$\mathbb{R}$	the real numbers
$\mathbb{Q}$	the rational numbers
$\mathbb{I}$	the irrational numbers
$\mathbb{Z}$	the integers
$\mathbb{P}$	the prime numbers

The symbols $\in$ and $\notin$, respectively, mean "is an element of" and "is not an element of." If x is an element of a set X, we write $x \in X$. If x is not an element of X, we write $x \notin X$. Sets are specified within curly brackets using "set builder notation." For example,

$$X = \{1, 2, 3\}$$

says that "X is the set containing elements 1, 2, and 3." Here, because 1 is an element of X, it is true that $1 \in X$. If the set is too long to enumerate, ellipses may be used. For example,

$$X = \{1, 2, 3, \ldots, 1{,}000\}$$

is the set of all integers from 1 to 1,000. An ellipsis adjacent to a curly bracket indicates continuation to infinity. For example,

$$X = \{1, 2, 3, \ldots\}$$

is the set of all positive integers. The unique set having no elements, called the *null set* or *empty set*, is associated with the symbol $\emptyset$:

$$\emptyset = \{\ \}$$

Sets also can be constructed by the specification of criteria. In general, the form looks like this:

$$X = \{\text{formula for candidates} \mid \text{criteria}\}$$

A colon can also be used instead of a vertical bar; they are interchangeable as separators:

$$X = \{\text{formula for candidates} : \text{criteria}\}$$

This general form can be read aloud like this: "X is the set of every candidate such that the candidate meets the criteria for membership." The criteria consist of at least one predicate. Let's look at some examples. Consider this set:

$$X = \{x \text{ is an animal} \mid x \text{ can fly}\}$$

Here, X is the set of all animals that can fly. The candidates are all animals; the criterion for membership is the ability to fly. Standard formal systems of mathematics require the candidates to exist in some predefined set (here, the set of all animals) to avoid paradoxical constructions. Here are some more mathematical examples:

Set	Interpretation
$\{x \in \mathbb{N} \mid x \leq 100\}$	all natural numbers up to and including 100
$\{x \in \mathbb{N} \mid \frac{x}{2} \in \mathbb{N}\}$	all even natural numbers
$\{x \in \mathbb{I} \mid x \neq \pi\}$	all irrational numbers except pi
$\{x \in \mathbb{R} \mid x \notin \mathbb{Q}\} = \mathbb{I}$	all real numbers that are not rational
$\{x^2 \in \mathbb{R} \mid x \in \mathbb{Z}\}$	all perfect squares

A *subset* Y of some set X is a set such that every element of Y is also an element of X. The symbol for "is a subset of" is $\subseteq$. The expression $Y \subseteq X$ is interpreted as "Y is a subset of X." The converse, "is not a subset of," is represented by the symbol $\nsubseteq$. For example:

$$\{1, 2\} \subseteq \{1, 2, 3, 4\}$$

$$\{1, 5\} \nsubseteq \{1, 2, 3, 4\}$$

The *union* of two sets is the set generated by collecting all elements that are in either of the original sets. The union symbol is $\cup$. Hence $X \cup Y$, the union of X and Y, is the set of all elements that are in either X or Y. Here's a mnemonic: *Union* starts with the letter U, which looks like the union symbol $\cup$. An example:

$$\{1, 2\} \cup \{3, 4\} = \{1, 2, 3, 4\}$$

The *intersection* of two sets is the set generated by collecting all the elements that are in both of the original sets. The intersection symbol is $\cap$. Hence, $X \cap Y$, the intersection of X and Y, is the set of all elements that are in both X and Y. Here's a mnemonic: Intersection entails the AND operator, which begins with the letter A, whose shape resembles the intersection symbol $\cap$. An example:

$$\{1, 2, 3\} \cap \{3, 4, 5\} = \{3\}$$

The *power set* of X, or the set of all sets of X including the empty set and X itself, is denoted as $\mathcal{P}(X)$.

Intervals on the real number line are denoted using square brackets or parentheses, depending on whether or not the boundary points are included in the set. Square brackets indicate inclusion, and parentheses indicate exclusion. For example, $[0, 1)$ is the set of all real numbers from zero to 1, including zero but excluding 1. This interval is said to be *closed* on the left and *open* on the right.

Physics

Symbol	Meaning
c	speed of light
$\hbar$	reduced Planck's constant

About the Contributors

Nicholas Laurita, Ph.D.

Author of the Classical Physics Chapter

Dr. Laurita was born in Oceanside, New York, and raised in Lakeland, Florida. He obtained a BSc degree in applied physics from the University of South Florida as a Presidential Scholar, graduating *summa cum laude*. He did his graduate research in the physics Ph.D. program at Johns Hopkins University, where he focused on examining the low-energy electrodynamics of quantum magnets. While a graduate student, he was awarded an Owen Scholars Fellowship, received the Rowland Prize for innovation and excellence in teaching, and was named a two-time National Achievement Rewards for College Scientists Scholar. In 2017, Dr. Laurita joined the Institute of Quantum Information and Matter (IQIM) at the California Institute of Technology as an IQIM Postdoctoral Fellowship recipient. His research continues to focus on using light to probe the electrodynamics of quantum matter. He has published multiple papers in reputable peer-reviewed journals and spoken at conferences around the world. He currently resides in Los Angeles with his wife, Nicole, and dog, Pepper.

Aidan Chatwin-Davies, Ph.D.

Author of the Special Relativity Chapter

Coauthor of the Quantum Mechanics Chapter

Dr. Chatwin-Davies is a Canadian-French theoretical physicist. He received bachelor's and master's degrees in applied mathematics from the University of Waterloo, and he received a doctorate in physics from the California Institute of Technology in 2018. He works on questions where gravity and quantum information intersect, which gives him the chance to think about problems in areas as diverse as cosmology, holography, emergent spacetime, and black hole physics. He is currently a postdoctoral fellow at the Institute for Theoretical Physics at KU Leuven in Belgium. Outside of physics, he enjoys playing the clarinet, getting better at speaking and reading Japanese, and backpacking in the wilderness.

Michael Coughlin, Ph.D.

Coauthor of the Quantum Mechanics Chapter

Dr. Coughlin graduated *summa cum laude* from Carleton College in Northfield, Minnesota, with a double major in physics and mathematics in 2012. He was awarded the Churchill Scholarship for study at the University of Cambridge, receiving an MPhil in astronomy in 2013. In September 2016, Dr. Coughlin received his physics Ph.D. at Harvard University, working with Professor Christopher Stubbs, with a thesis entitled "Gravitational-wave astronomy in the LSST era." He is currently the David and Ellen Lee Prize Postdoctoral Fellow in physics, math, and astronomy at the California Institute of Technology. He has been a member of the LIGO Scientific Collaboration for almost a decade, and more recently, he has been involved in current and planned telescope surveys such as the Large Synoptic Survey Telescope (LSST), Zwicky Transient Facility (ZTF), Panoramic Survey Telescope and Rapid Response System (Pan-STARRS), and Asteroid Terrestrial-impact Last Alert System (ATLAS), working at the interface of gravitational-wave and time-domain astronomy. He is a coauthor of more than a hundred publications, including more than twenty as lead or second author. Outside of science, he dances on the Caltech ballroom dance team and volunteers at science nights at local elementary schools with the LIGO education and public outreach team.

Grant Sanderson

Contributing Essayist and Poet

Grant Sanderson is the creator of 3blue1brown, a popular YouTube channel with a community of over 1 million subscribers and 45 million views. The goal of the channel is to create more mathematics aficionados. The videos cover a variety of topics in math, ranging from student-focused material, like series on calculus and linear algebra, to topics outside standard academic curricula. The channel also includes material on related fields, such as neural networks, cryptocurrency, and quantum physics. Sanderson studied mathematics and computer science at Stanford University. After graduating, he began programming visualizations to explain challenging ideas in mathematics. He was selected in 2015 by Khan Academy's talent search and was eventually named the "multivariable calculus fellow." He stayed with Khan until the end of 2016 as a writer and video creator, all the while developing 3blue1brown.

Bibliography

Allis, Victor, and Teun Koetsier. (1991). "On Some Paradoxes of the Infinite." *British Journal for the Philosophy of Science*. 42 (2): 187–194.

Aristotle. *Topics*. 1.12 105a13–14.

Arrow, Kenneth. (1951). "Social Choice and Individual Values." *Social Choice and Individual Values*. New York: John Wiley & Sons.

Arrow, Kenneth. (1974). *The Limits of Organization*. New York: W. W. Norton and Company.

Associated Press. (2005). "Game Theorists Win Prize for Economics." *Los Angeles Daily News*. https://www.dailynews.com/2005/10/11/game-theorists-win-prize-for-econo mics/ (last updated August 29, 2017).

Banach, Stefan, and Albert Tarski. (1924). "Sur la décomposition des ensembles de points en parties respectivement congruentes." *Fundamenta Mathematicae* (in French). 6: 244–277.

Barbeau, E. J., and P. J. Leah. (1976). "Euler's 1760 Paper on Divergent Series." *Historia Mathematica*. 3: 141–160.

Barwise, Jon. (1975). *Admissible Sets and Structures*. Berlin: Springer-Verlag.

Barwise, Jon, and John Etchemendy. (1987). *The Liar: An Essay in Truth and Circularity*. New York: Oxford University Press.

Bell, John S., and Alain Aspect. (2004). *Speakable and Unspeakable in Quantum Mechanics: Collected Papers on Quantum Philosophy*. Cambridge: Cambridge University Press.

Benacerraf, Paul. (1962). "Tasks, Super-Tasks, and the Modern Eleatics." *Journal of Philosophy*. 59 (24): 765–784.

Bennett, Charles. (1982). "The Thermodynamics of Computation—A Review." *International Journal of Theoretical Physics*. 21 (12): 905–940.

Bennett, Charles. (1987). "Demons, Engines, and the Second Law." *Scientific American.* 257 (5): 108–116.

Bennett, Charles, and Benjamin Schumacher. (2011). "Maxwell's Demons Appear in the Lab." *Nikkei Science.* 3–6.

Bernoulli, Daniel. (1738). Trans. Louise Sommer (1954). "Exposition of a New Theory on the Measurement of Risk." *Econometrica.* 22 (1): 22–36.

Bertrand, Joseph. (1883). "Review of *Theorie mathematique de la richesse sociale* and of *Recherches sur les principles mathematiques de la theorie des richesses.*" *Journal des Savants.* 67: 499–508.

Bickel, P. J., E. A. Hammel, and J. W. O'Connell. (1975). "Sex Bias in Graduate Admissions: Data from Berkeley." *Science.* 187 (4175): 398–404.

Braess, Dietrich. (1968). "Über ein Paradoxon aus der Verkehrsplanung." *Unternehmensforschung.* 12: 258–268. Trans. Dietrich Braess, A. Nagurney, and T. Wakolbinger (2005). "On a Paradox of Traffic Planning." *Transportation Science.* 39: 446–450.

Brillouin, Léon. (1951). "Maxwell's Demon Cannot Operate: Information and Entropy. I." *Journal of Applied Physics.* 22 (3): 334–337.

Broome, John. (1995). "The Two-Envelope Paradox." *Analysis.* 55 (1): 6–11.

Cantor, Georg. (1874). "Über eine Eigenschaft des Inbegriffes aller reellen algebraischen Zahlen." *Journal für die reine und angewandte Mathematik.* 77: 258–262. In his *Gesammelte Abhandlungen*, 145–148. English translation by W. Ewald in Ewald (1996), Vol. 2, 840–843.

Carroll, Sean M. (2004). *Spacetime and Geometry: An Introduction to General Relativity.* San Francisco: Addison Wesley.

Carroll, Sean. (2014). *Spacetime and Geometry: An Introduction to General Relativity.* Pearson New International Edition. Essex, UK: Pearson.

Chalmers, David J. (2002). "The St. Petersburg Two-Envelope Paradox." *Analysis.* 62 (2): 155–157.

Chow, Timothy Y. (1998). "The Surprise Examination or Unexpected Hanging Paradox." *American Mathematicaly Monthly.* 105: 41–51.

Christensen, Ronald, and Jessica Utts. (1992). "Bayesian Resolution of the *Exchange Paradox.*" *The American Statistician.* 46 (4): 274–276.

Clauser, John F., Michael A. Horne, Abner Shimony, and Richard A. Holt. (1969). "Proposed Experiment to Test Local Hidden-Variable Theories." *Physical Review Letters.* 23 (15): 880. Erratum (1970) in *Physical Review Letters*, 24 (10): 549.

Cover, Thomas M. (1987). "Pick the Largest Number." *Open Problems in Communication and Computation.* Cover, Thomas M. Cover and B. Gopinath, eds. New York: Springer-Verlag.

Creutz, Edward C. (2005). "Feynman's Reverse Sprinkler." *American Journal of Physics.* 73 (3): 198–199.

Curry, Haskell. (1942). "The Inconsistency of Certain Formal Logics." *Journal of Symbolic Logic.* 7 (3): 115–117.

Da Vinci, Leonardo, and Edward McCurdy. (2009). *The Notebooks of Leonardo da Vinci.* London: Folio Society.

Deutsch, David. (1997). *The Fabric of Reality.* New York: Penguin Books.

Deutsch, David. (2011). *The Beginning of Infinity.* New York: Viking Press.

Dewan, Edmond, and Michael Beran. (1959). "Note on Stress Effects due to Relativistic Contraction." *American Journal of Physics.* 27 (7): 517–518.

Dunham, William. (1990). *Journey Through Genius: The Great Theorems of Mathematics.* New York: John Wiley & Sons.

Ehrenfest, Paul. (1909). "Gleichförmige Rotation starrer Körper und Relativitätstheorie." *Physikalische Zeitschrift.* 10: 918.

Einstein, Albert. (1905). "Zur Elektrodynamik bewegter Körper." *Annalen der Physik.* 17: 891–921. [*Annalen der Physik* 14: 194–224 (2005)].

Einstein, Albert. (1920). *Relativity.* Amherst, NY: Prometheus Books.

Einstein, Albert. (1935, May 5). "Letter to the Editor." [Obituary for Emmy Noether.] *New York Times.*

Einstein, Albert, Boris Podolsky, and Nathan Rosen. (1935). "Can Quantum-Mechanical Description of Physical Reality Be Considered Complete?" *Physical Review.* 47 (10): 777–780.

Eldridge-Smith, Peter. (2011). "Pinocchio against the Dialetheists." *Analysis.* 71 (2): 306–308.

Elga, Adam. (2000). "Self-Locating Belief and the Sleeping Beauty Problem." *Analysis.* 60 (2): 143–147.

Euler, Leonhard. *De seriebus divergentibus, Opera Omnia.* 1, 14, 585–617.

Falk, Ruma. (2008). "The Unrelenting Exchange Paradox." *Teaching Statistics.* 30 (3): 86–88.

Falk, Ruma, and Raymond Nickerson. (2009). "An Inside Look at the Two Envelopes Paradox." *Teaching Statistics.* 31 (2): 39–41.

Feynman, Richard. (1982). "Simulating Physics with Computers." *International Journal of Theoretical Physics*. 21 (6–7): 467–488.

Feynman, Richard. (1985). *Surely You're Joking, Mr. Feynman!* New York: W. W. Norton and Company.

Feynman, Richard Phillips, Robert B. Leighton, and Matthew Linzee Sands. (1963). *The Feynman Lectures on Physics*. Boston: Addison-Wesley.

Fitch, F. A. (1964). "A Goedelized Formulation of the Prediction Paradox." *American Philosophical Quarterly*. 1 (2): 161–164.

Frege, Gottlob. (1902). "Letter to Russell." In Jean van Heijenoort (1967), *From Frege to Gödel*. Cambridge, MA: Harvard University Press (pp. 126–128).

Gabor, Dennis. (1961). "Light and Information." *Progress in Optics*. 1, 111–152.

Gardner, Martin. (1959, October). "Mathematical Games." *Scientific American*. 180–182.

Gibbard, Allan. (1973). "Manipulation of Voting Schemes: A General Result." *Econometrica*. 41 (4): 587–601.

Gintis, Herbert. (2000). *Game Theory Evolving*. Princeton, NJ: Princeton University Press.

Gleick, James. (1992). *Genius: The Life and Science of Richard Feynman*. New York: Pantheon Books.

Gödel, Kurt. (1931). "Über formal unentscheidbare Sätze der Principia Mathematica und verwandter Systeme I." *Monatshefte für Mathematik Physik*. 38: 173–198.

Good, I. J. (1960). "The Paradox of Confirmation." *British Journal for the Philosophy of Science*. 11 (42): 145–149.

Good, I. J. (1967). "The White Shoe Is a Red Herring." *British Journal for the Philosophy of Science*. 17 (4): 322.

Griffiths, David J. (2004). *Introduction to Quantum Mechanics*. 2nd ed. Essex, UK: Pearson.

Grøn, Øyvind. (1975). "Relativistic Description of a Rotating Disk." *American Journal of Physics*. 43 (10): 869–876.

Hardy, G. H. (1949). *Divergent Series*. Oxford: Oxford University Press.

Harmer, Gregory, Derek Abbott, and Peter Taylor. (2000). "The Paradox of Parrondo's Games." *Royal Society's Proceedings: Mathematical, Physical, and Engineering Sciences*. 456 (1994): 247–259.

Hausdorff, Felix. (1919). "Dimension und äußeres Maß." *Mathematische Annalen.* 79 (1-2): 157–179.

Helliwell, Thomas M. (2010). *Special Relativity.* Sausalito, CA: University Science Books.

Hempel, C. G. (1937). "Le problème de la vérité." *Theoria.* 3 (2-3): 206–244.

Hempel, C. G. (1943). "A Purely Syntactical Definition of Confirmation." *Journal of Symbolic Logic.* 8 (4): 122–143.

Hempel, C. G. (1945). "Studies in the Logic of Confirmation." *Mind.* 54 (213): 1–26.

Hertzberg, Hendrik. (2010, February 15). "And the Oscar Goes To." *The New Yorker.*

Hilbert, David, William Ewald, and Wilfriend Sieg. (2013). "David Hilbert's Lectures on the Foundations of Arithmetics and Logic 1917–1933." Heidelberg: Springer-Verlag.

Hofstadter, Douglas. (1979). *Gödel, Escher, Bach: An Eternal Golden Braid.* New York: Basic Books.

Hoogeveen, Joost, Szczepan Kowalczyk, and Maarten van de Meent. (2004). "An Historical and Modern View on Bell's Inequality." Retrieved from https://pdfs.semanticscholar.org/1a8f/1fa3f22a11e3ea437000ced343d26\2ed5b4b.pdf.

Hughes, G. E. (1992). *John Buridan on Self-Reference: Chapter Eight of Buridan's Sophismata, with a Translation, and Introduction, and a Philosophical Commentary.* Cambridge: Cambridge University Press.

Jenkins, Alejandro. (2004). "An Elementary Treatment of the Reverse Sprinkler." *American Journal of Physics.* 72 (10): 1276–1282.

Jenkins, Alejandro. (2011). "Sprinkler Head Revisited: Momentum, Forces, and Flows in Machian Propulsion." *European Journal of Physics.* 32 (5): 1213–1226.

Kaplan, David, and Richard Montague. (1960). "A Paradox Regained." *Notre Dame Journal of Formal Logic.* 1: 79–90.

Kardar, Mehran. (2007). *Statistical Physics of Particles.* Cambridge: Cambridge University Press.

Kirwan, Christopher. (1924). Translation with notes. *Aristotle's Metaphysics Books Gamma, Delta and Epsilon.* 2nd ed. Oxford: Clarendon Press.

Kittel, Charles, and Herbert Kroemer. (1980). *Thermal Physics.* 2nd ed. New York: W. H. Freeman and Company.

Koch, Helge von. (1904). "Sur une courbe continue sans tangente, obtenue par une construction géométrique élémentaire." *Arkiv för matematik, astronomi och fysik.* 1: 681–702.

Kraitchik, Maurice. (1953). *Mathematical Recreations*. Revised ed. Mineola, NY: Dover Publications.

Kripke, Saul. (1975). "An Outline of a Theory of Truth." *Journal of Philosophy*. 72: 690–716.

Landauer, Rolf. (1961). "Irreversibility and Heat Generation in the Computing Process." *IBM Journal of Research and Development*. 5 (3): 183–191.

Laraudogoitia, Pérez J. (1996). "A Beautiful Supertask." *Mind*. 105 (417): 81–83.

Laraudogoitia, Pérez J. (1997). "Classical Particle Dynamics, Indeterminism, and a Supertask." *British Journal for the Philosophy of Science*. 48 (1): 49–54.

Lefebvre, Neil, and Melissa Schehlein. (2005). "The Liar Lied." *Philosophy Now*. Issue 51.

Littlewood, John Edensor. (1953). *A Mathematician's Miscellany*. London: Methuen.

Mach, Ernst. (1883). Trans. Thomas Joseph McCormack (1915). *The Science of Mechanics: A Critical and Historical Account of its Development*. Chicago: Open Court Publishing Company.

Madigan, Arthur S. J. (1993). Translation. *Alexander of Aphrodisias: On Aristotle's Metaphysics 4*, with appendix by Richard Sorabji. Ithaca, NY: Cornell University Press.

Maher, P. (1999). "Inductive Logic and the Ravens Paradox." *Philosophy of Science*. 66 (1): 50–70.

Maldacena, Juan, and Leonard Susskind. (2013). "Cool Horizons for Entangled Black Holes." *Progress of Physics*. 61 (9): 781–811.

Mandelbrot, Benoit. (1982). *The Fractal Geometry of Nature*. New York: W. H. Freeman and Company.

Martin, Robert, and Peter Woodruff. (1975). "On Representing 'True-in-L' in L." *Philosophia*. 5(3): 217–221.

Maxwell, James Clerk. (1872). *Theory of Heat*. London: Longmans, Green, and Co.

McCaskey, John. (2014). "Induction in the Socratic Tradition." In Paolo C. Biondi and Louis Groarke, eds., *Shifting the Paradigm: Alternative Perspectives in Induction*, 161–192. De Gruyter.

McCaskey, John. (2015, March 12). "The History of Induction." [Blog post]. Retrieved from http://www.johnmccaskey.com/history-of-induction/.

Menger, Karl. (1926). "Allgemeine Räume und Cartesische Räume. I." *Communications to the Amsterdam Academy of Sciences*. English translation in Gerald A. Edgar, ed. (2004), *Classics on Fractals, Studies in Nonlinearity*. Boulder, CO: Westview Press, Advanced Book Program.

Merzbacher, Eugen. (1997). *Quantum Mechanics*. 3rd ed. New York: John Wiley & Sons.

Michelson, Albert A., and Edward W. Morley. (1887). "On the Relative Motion of the Earth and the Luminiferous Ether." *American Journal of Science*. 34 (203): 333–345.

Mills, Eugene. (1998). "A Simple Solution to the Liar." *Philosophical Studies*. 89 (2–3): 197–212.

de Montmort, Pierre Ramond. (1713). "Essay d'analyse sur les jeux de hazard." (2nd ed., in French). Providence, RI: American Mathematical Society.

Nalebuff, Barry. (1989). "Puzzles: The Other Person's Envelope Is Always Greener." *Journal of Economic Perspectives*. 3 (1): 171–181.

Nash, John. (1950). "Equilibrium Points in n-Person Games." *Proceedings of the National Academy of Sciences of the United States of America*. 36 (1): 48–49.

O'Connor, D. J. (1948). "Pragmatic Paradoxes." *Mind*. 57: 358–359.

Ohanian, Hans and Remo Ruffini. (1994). *Gravitation and Spacetime*. 2nd ed. New York: W. W. Norton & Company.

Parrondo, Juan, and Pep Español. (1996). "Criticism of Feynman's Analysis of the Ratchet as an Engine." *American Journal of Physics*. 64: 1125–1130.

Pauli, Wolfgang. (1958). *Theory of Relativity*. New York: Pergamon Press.

Piccione, Michele, and Ariel Rubinstein. (1997). "On the Interpretation of Decision Problems with Imperfect Recall." *Games and Economic Behavior*. 19: 3–24.

Pólya, George. (1954). *Induction and Analogy in Mathematics*. Vol. 1 of *Mathematics and Plausible Reasoning*. Princeton, NJ: Princeton University Press.

Powers, Michael R. (2015). "Paradox-Proof Utility Functions for Heavy-Tailed Payoffs: Two Instructive Two-Envelope Problems." *Risks*. 3 (1): 26–34.

Priest, Graham. (1979). "The Logic of Paradox." *Journal of Philosophical Logic*. 8: 219–241.

Priest, Graham. (1984). "Logic of Paradox Revisited." *Journal of Philosophical Logic*. 13: 153–179.

Prior, A. N. (1958). "Epimenides the Cretan." *Journal of Symbolic Logic*. 23: 261–266.

Prior, A. N. (1961). "On a Family of Paradoxes." *Notre Dame Journal of Formal Logic*. 2: 16–32.

Prior, A. N. (1976). *Papers in Logic and Ethics*. London: Gerald Duckworth & Co.

Pulskamp, Richard J. (2013). "Correspondence of Nicolas Bernoulli Concerning the St. Petersburg Game." Trans. *Die Werke von Jakob Bernoulli*, Band 3, K9, as published by Birkhäuser Basel.

Quine, W. V. O. (1976). "The Ways of Paradox." From *The Ways of Paradox and Other Essays*, revised ed. Cambridge, MA: Harvard University Press.

Rand, Ayn. (1957). *Atlas Shrugged*. New York: Random House.

Rand, Ayn. (1964). *The Virtue of Selfishness*. New York: New American Library.

Rand, Ayn. (1982). *Philosophy: Who Needs It?*. Indianapolis: Bobbs-Merrill.

Rand, Ayn. (1990). *Introduction to Objectivist Epistemology*. New York: New American Library.

Rizzi, Guido, and Matteo L. Ruggiero. (2002). "Space Geometry of Rotating Platforms: An Operational Approach." *Foundations of Physics*. 32 (10): 1525–1556.

Romero, Gustavo. (2014). "The Collapse of Supertasks." *Foundations of Science*. 19 (2): 209–216.

Rosenthal, Robert. (1981). "Games of Perfect Information, Predatory Pricing and the Chain-Store Paradox." *Journal of Economic Theory*. 25 (1): 92–100.

Rubinstein, Ariel. (1989). "The Electronic Mail Game: Strategic Behavior Under 'Almost Common Knowledge.'" *American Economic Review*. 79 (3): 385–391.

Russell, Bertrand. (1903). *The Principles of Mathematics*, Vol. 1. Cambridge: Cambridge University Press.

Russell, Bertrand. (1908). "Mathematical Logic as Based on the Theory of Types." *American Journal of Mathematics*. 30 (3): 222–262.

Russell, Bertrand. (1914). *Our Knowledge of the External World*. Chicago: Open Court Publishing Company.

Russell, Bertrand. (1919). *Introduction to Mathematical Philosophy*. London: George Allen and Unwin.

Russell, Bertrand, and Alfred Whitehead. (1910). *Principia Mathematica*, Volume 1. Cambridge: Cambridge University Press.

Russell, Bertrand, and Alfred Whitehead. (1912). *Principia Mathematica*, Vol. 2. Cambridge: Cambridge University Press.

Russell, Bertrand, and Alfred Whitehead. (1913). *Principia Mathematica*, Vol. 3. Cambridge: Cambridge University Press.

Satterthwaite, Mark Allen. (1975). "Strategy-Proofness and Arrow's Conditions: Existence and Correspondence Theorems for Voting Procedures and Social Welfare Functions." *Journal of Economic Theory*. 10 (2): 187–217.

Schultz, A. K. (1987). "Comment on the Inverse Sprinkler Problem." *American Journal of Physics*. 55 (6): 488–489.

Selvin, Steve. (1975a). "On the Monty Hall Problem (Letter to the Editor)." *American Statistician*. 29 (3): 134.

Selvin, Steve. (1975b). "A Problem in Probability (Letter to the Editor)." *American Statistician*. 29 (1): 67.

Serway, Raymond A., and John W. Jewett Jr. (2008). *Physics For Scientists and Engineers*. 7th ed. Pacific Grove, CA: Thomson Brooks/Cole.

Sierpiński, Waclaw. (1916). "On Curves Which Contain the Image of Any Given Curve." *Matematicheskii Sbornik*. 30: 267–287. Reprinted in *Oeuvres Choisies*, Vol. 2, 107–119.

Simanek, Donald E. (2012). "Perpetual Futility: A Short History of the Search for Perpetual Motion." Retrieved from https://www.lockhaven.\edu/~dsimanek/museum /people/people.htm.

Simpson, Edward. (1951). "The Interpretation of Interaction in Contingency Tables." *Journal of the Royal Statistical Society, Series B*. 13: 238–241.

Smith, Henry John Stephen (1874). "On the Integration of Discontinuous Functions." *Proceedings of the London Mathematical Society*. 1 (6): 140–153.

Smoluchowski, Marion. (1912). "Experimentell nachweisbare, der üblichen Thermodynamik widersprechende Molekularphänomene." *Physikalische Zeitschrift*. 8: 1069–1080.

Smullyan, Raymond. (1992). *Satan, Cantor, and Infinity and Other Mind–Boggling Puzzles*. New York: Alfred A. Knopf.

Stevenson, Robert Louis. (1891). "The Bottle Imp." In *South Sea Tales* (1996). Oxford: Oxford University Press.

Szabó, László E. (2007). "The Einstein-Podolsky-Rosen Argument and the Bell Inequalities." arXiv:0712.1318 [quant-ph].

Szilárd, Leó. (1929). "On the Decrease of Entropy in a Thermodynamic System by the Intervention of Intelligent Beings." *Zeitschrift für Physik*. 53: 840–856.

Tarski, Alfred. (1935). "Der Wahrheitsbegriff in den formalisierten Sprachen." *Studia Philosophica*. 1: 261–405.

Tarski, Alfred. (1944). "The Semantic Conception of Truth and the Foundations of Semantics." *Philosophy and Phenomenological Research*. 4 (3): 341–376.

Tarski, Alfred. (1956). *Logic, Semantics, Metamathematics: Papers from 1923–1938*. Oxford: Clarendon Press.

Tarski, Alfred. (1994). *Introduction to Logic and to the Methodology of the Deductive Sciences*. New York: Oxford University Press. Originally published in Polish in 1936.

Thomson, James F. (1954). "Tasks and Super-Tasks." *Analysis*. 15 (1): 1–13.

Thomson, William. (1874). "Kinetic Theory of the Dissipation of Energy." *Nature*. 9 (232): 441–444.

Townsend, John. (2012). *A Modern Approach to Quantum Mechanics*. Second edition. Sausalito, CA: University Science Books.

Vieira, Ricardo S. (2017). "Solution of Supplee's Submarine Paradox Through Special and General Relativity." *Europhysics Letters*. 116 (5): 50007p1-7.

von Neumann, John, and Oskar Morgenstern. (1944). *Theory of Games and Economic Behavior*. Princeton, NJ: Princeton University Press.

vos Savant, Marilyn. (1990, September 9). "Ask Marilyn." *Parade Magazine*. 16.

Weinberg, Steven. (2012). *Lectures on Quantum Mechanics*. Cambridge: Cambridge University Press.

Weiss, Michael. (2017). "The Rotating Disk in Relativity." Retrieved from http://math .ucr.edu/home/baez/physics/Relativity/SR/rigid_disk.ht\ml.

Weiss, Michael, and Don Koks. (2017). "Bell's Spaceship Paradox." Retrieved from http://math.ucr.edu/home/baez/physics/Relativity/SR/BellSpac\eships/spaceship_ puzzle.html.

Yablo, Stephen. (1993a). "Hop, Skip, and Jump: The Agonistic Conception of Truth." *Philosophical Perspectives*. 3: 371–396.

Yablo, Stephen. (1993b). "Paradox Without Self-Reference." *Analysis*. 53 (4): 251–252.

Zuboff, Arnold. (1990). "One Self: The Logic of Experience." *Inquiry*. 33 (1): 39–68.

Index